T0205649

FOUNDATIONS OF NANOTECHNOLOGY

VOLUME 1

PORE SIZE IN CARBON-BASED NANO-ADSORBENTS

FOUNDATIONS OF NANOTECHNOLOGY

VOLUME 1

PORE SIZE IN CARBON-BASED NANO-ADSORBENTS

A. K. Haghi, PhD, Sabu Thomas, PhD, and
Moein MehdiPour MirMahaleh

Apple Academic Press

TORONTO   NEW JERSEY

Apple Academic Press Inc. | Apple Academic Press Inc.
3333 Mistwell Crescent | 9 Spinnaker Way
Oakville, ON L6L 0A2 | Waretown, NJ 08758
Canada | USA

©2015 by Apple Academic Press, Inc.

First issued in paperback 2021

Exclusive worldwide distribution by CRC Press, a member of Taylor & Francis Group
No claim to original U.S. Government works

ISBN 13: 978-1-77463-104-1 (pbk)
ISBN 13: 978-1-77188-026-8 (hbk)

Library of Congress Control Number: 2014946904

Library and Archives Canada Cataloguing in Publication

Foundations of nanotechnology.
(AAP research notes on nanoscience & nanotechnology book series)
Contents: Volume 1. Pore size in carbon-based nano-adsorbents/A.K. Haghi, PhD, Sabu Thomas, PhD, and Moein MehdiPour MirMahaleh

Includes bibliographical references and index.
ISBN 978-1-77188-026-8 (v. 1: bound)
1. Nanotechnology. I. Series: AAP research notes on nanoscience & nanotechnology book series

| T174.7.F69 2014 | 620'.5 | C2014-905376-2 |

Apple Academic Press also publishes its books in a variety of electronic formats. Some content that appears in print may not be available in electronic format. For information about Apple Academic Press products, visit our website at **www.appleacademicpress.com** and the CRC Press website at **www.crcpress.com**

ABOUT AAP RESEARCH NOTES ON NANOSCIENCE & NANOTECHNOLOGY

BOOKS IN THE AAP RESEARCH NOTES ON NANOSCIENCE & NANOTECHNOLOGY BOOK SERIES

Nanostructure, Nanosystems and Nanostructured Materials:
Theory, Production, and Development
Editors: P. M. Sivakumar, PhD, Vladimir I. Kodolov, DSc,
Gennady E. Zaikov, DSc, and A. K. Haghi, PhD

Nanostructures, Nanomaterials, and Nanotechnologies to Nanoindustry
Editors: Vladimir I. Kodolov, DSc, Gennady E. Zaikov, DSc, and
A. K. Haghi, PhD

Foundations of Nanotechnology:
Volume 1: Pore Size in Carbon-Based Nano-Adsorbents
A. K. Haghi, PhD, Sabu Thomas, PhD, and Moein MehdiPour MirMahaleh

Foundations of Nanotechnology:
Volume 2: Nanoelements Formation and Interaction
Sabu Thomas, PhD, Saeedeh Rafiei, Shima Maghsoodlou, and Arezo Afzali

Foundations of Nanotechnology:
Volume 3: Mechanics of Carbon Nanotubes
Saeedeh Rafiei

ABOUT THE AUTHORS

A. K. Haghi, PhD

A. K. Haghi, PhD, holds a BSc in urban and environmental engineering from University of North Carolina (USA); a MSc in mechanical engineering from North Carolina A&T State University (USA); a DEA in applied mechanics, acoustics and materials from Université de Technologie de Compiègne (France); and a PhD in engineering sciences from Université de Franche-Comté (France). He is the author and editor of 65 books as well as 1000 published papers in various journals and conference proceedings. Dr. Haghi has received several grants, consulted for a number of major corporations, and is a frequent speaker to national and international audiences. Since 1983, he served as a professor at several universities. He is currently Editor-in-Chief of the *International Journal of Chemoinformatics and Chemical Engineering* and *Polymers Research Journal* and on the editorial boards of many international journals. He is a member of the Canadian Research and Development Center of Sciences and Cultures (CRDCSC), Montreal, Quebec, Canada

Sabu Thomas, PhD

Dr. Sabu Thomas is the Director of the School of Chemical Sciences, Mahatma Gandhi University, Kottayam, India. He is also a full professor of polymer science and engineering and the Director of the International and Inter University Centre for Nanoscience and Nanotechnology of the same university. He is a fellow of many professional bodies. Professor Thomas has authored or co-authored many papers in international peer-reviewed journals in the area of polymer processing. He has organized several international conferences and has more than 420 publications, 11 books and two patents to his credit. He has been involved in a number of books both as author and editor. He is a reviewer to many international journals and has received many awards for his excellent work in polymer processing. His h Index is 42. Professor Thomas is listed as the 5th position in the list of Most Productive Researchers in India, in 2008.

Moein MehdiPour MirMahaleh

Moein MehdiPour MirMahaleh is a professional textile engineer and is a Research Fellow at Technopark, Kerala, India's first technology park and among the three largest IT parks in India today.

CONTENTS

LIST OF ABBREVIATIONS

ACF	Activated Carbon Fibers
AFM	Atomic Force Microscopy
ASA	Adsorption Stochastic Algorithm (for simulation)
ATRP	Atom Transfer Radical Polymerization
BET	Brunauer-Emmet-Teller Method
CCVD	Catalytic Chemical Vapor Deposition
CNT	Carbon Nanotube
CNT/MO	Carbon Nanotubes/Metal Oxide
CPMD	Car–Parrinello Molecular Dynamics
CRTs	Cathode Ray Tubes
CTAB	CetylTrimethylammonium Bromide
CVD	Chemical Vapor Decomposition
CVI	Chemical Vapour Infiltration
DA	Dubinin-Astakhov Model
DC	Determination Coefficient
DC	Direct Current
DCV-GCMD	Dual-Control-Volume of MDS
DFT	Density Functional Theory
DFT	Density Functional Theory
DOM	Dissolved Natural Organic Matter
DRS	Dubinin-Radushkevich-Stoeckly Model
DS	Dubinin-Stoeckly Model
EDLC	Electric Double-Layer Capacitor
EDLCs	Electrodes Electric Double Layer Capacitors
erf(W)	Error Function of W Parameter
FEDs	Field Emission Displays
GCMC	Grand Canonical Monte Carlo
GCMS	Grand Canonical Monte Carlo simulations
HC	Hydrocarbons
HFPECVD	Hot Filament Plasma Enhanced Chemical Vapour Deposition
HK	Horvath-Kawazoe Model
HOPG	High Pyrolitic Graphite
HREM	High-Resolution Electron Microscopy
IHK	Improved Horvath-Kawazoe model
ITO	Indium Tin Oxide

IUPAC	International Union for Pure and Applied Chemistry
LJ	Lennard-Jones Potential Function
MDS	Molecular Dynamics Simulation
MSC	Molecular Sieving Carbon
MSD	Micropore Size Distribution
MWCNTs	Multi Walled Carbon Nanotubes
ND	The Nguyen and Do Method
NLDFT	Non Local Density Functional Theory
OGMT	Oil Gas Motherboard Thicknesses
OM	Organic Materials
PA	Polyamide
PC	Polycarbonate
PECVD	Plasma Chemical Vapour Deposition
PSA	Pressure Swing Adsorption
PSD	Pore Size Distribution
PVA	Polyvinyl Acetate
PVP	Polyvinylpyridine
SANS	Small-angle Neutron Scattering
SAXS	Small-angle X-ray Scattering
S_{BET}	Surface Area of BET Method
SDS	Sodiumdodecyl Sulfate
SOCs	Synthetic Organic Chemicals
STM	Scanning Tunneling Microscopy
SWCNTs	Single Walled Carbon Nanotubes
T&O	Taste and Odor
TEM	Transmission Electron Microscopy
TEP	Thermoelectric Power
TSA	Temperature Swing Adsorption
XRD	X-ray Diffraction

LIST OF SYMBOLS

$(L-d_a)$	The effective pore width
$(L-d_a)/d_A$	The reduced effective pore width
$\Pi_1 \& \Pi_2$	Characterize the strength of the surface forces field
$(L-d_a)_{p,max}$	The effective pore width related to the maximum value of function $J(L-d_a)/d_A)$
$(L-d_a)/d_A)_{p,max}$	The reduced effective pore width related to the maximum value of function $J(L-d_a)/d_A)$
A_{pot}	Free energy of adsorption in Dubinin–Astakhov isotherm equation) $(=-\Delta G^{ads})$
A_{HK}	Constant of HK method
$A_{max}, \bar{A} \& \sigma_A$	Proportional to the parameter ρ
C_1, C_2, C_3, C_4	Constants for a given adsorbate-adsorbent system
C_{BET}	BET constant on flat surface
$D_{in} \& \chi(D_{in})$	The normalized differential PSD function
$L_{av,I}$	The average effective width of the primary porous structure (L <~ 1)
$L_{av,II}$	The average effective width of the secondary porous structure (~1<L <~ 2) determined from DFT and/or ND
$L_{av,tot}$	The average effective width of the primary and secondary porous structure (L<~ 2)
L_{av}	The average pore width
N_{DRS}	The values of adsorption
N_{mDRS}	The values of maximum adsorption
P_e	Permeability of the membrane
S_{DFT}	The surface of pores calculated for all the pores
S_{DFT}^{meso}	The surface of pores calculated for all the pores with diameters larger and/or equal to 2 nm
S_{DFT}^{micro}	The surface of pores calculated for all the pores with diameters smaller and/or equal to 2 nm

$S_{c,\alpha}$	Total specific surface area
S^{diff}	The differential molar entropies of adsorbed molecules
S_g	The molar entropy of gas
S_{liq}	The molar entropy of liquid
$S_{me,\alpha}$	The specific surface area of mesopores calculated from the high-resolution αs-plot
S_{sol}	The molar entropy of solid
$V_{DFT,I}^{micro}$	The volume of the primary micropores determined from DFT
V_{DFT}^{meso}	The volume of the mesopores with diameters larger than and/or equal to 2 nm determined from DFT
V_{DFT}^{micro}	The total volume of the micropores with diameters smaller than and/or equal to 2 nm
V_{HK}	The micropore volumes determined from HK
V_{in}	Molar volume of the adsorbent
$V_{mi,\alpha}$	The volume of micropores calculated from the high-resolution αs plot
W_0	The volume of micropores calculated from the Dubinin–Astakhov and/or Dubinin–Radushkevich isotherm equation
$W_{exp.tot}$	The maximum value of the experimental adsorption
$X_{mi}(A)$	The adsorption potential distribution in micropores
α_{mi}	Equilibrium amount adsorbed in micropores
C_s	Complete filling concentration
d_A	Diameter of the adsorbent atom
d_a	Diameter of the adsorbate molecule
f_d	The downstream fugacities
f_u	The upstream fugacities respectively
p_0	The standard state pressure
p_s	Saturation pressure of an adsorbate
q_{diff}	The differential heat of adsorption
$z_{max}, \bar{z}$ & σ_z	Used to characterize the structural heterogeneity of microporous solids
γ_∞	Surface tension of the bulk fluid

γ_{me}	Amount adsorbed at relative pressure per unit area of the mesopore surface
γ_s	Amount adsorbed at relative pressure per unit area of nonporous reference adsorbent surface
ε_0	Adsorption characterized energy of standard state
$\lambda_1 \ \& \ \lambda_2$	Range of the structural forces action
ρ_b^*	The reduced number density of the bulk fluid
ρ_p^*	The reduced number density of the pore fluid
χ_{normDRS}	The pore size distribution was calculated using the correct normalization factor
ΔG^{ads}	Free energy of adsorption equal to $RT \ln(p/p_s)$
ΔH^{ads}	The enthalpy of adsorption
ΔH^{vap}	The enthalpy of vaporization
ΔS^{ads}	The entropy of adsorption
A	Average adsorption potential
a	Total amount adsorbed
a^0_{mes}	Monolayer capacity of the mesopore surface
a^0_{mi}	Maximum amount adsorbed in the micropores
a^0_s	Monolayer capacity of reference adsorbent from standard adsorption isotherm
a_m	Monolayer capacity
a_{mes}	Amount adsorbed on the mesopore surface
a_{mic}	Amount adsorbed on the micropore surface
$a_s(0.4)$	Amount adsorbed on the surface of reference solid at pressure 0.4
$a_s(p/p0)$	Amount adsorbed on the surface of reference adsorbent
B	Temperature-independent structural parameter of micropore sizes
C	Constant the heat of first layer adsorption
D_0	Constant Number
dA/dx	Derivative of adsorption potential
D_{max}, D_{min}	Minimum and Maximum pore size in the kernel of NLDFT
E_0	Characteristic energy of adsorption for the reference vapor
F	Conversion factor for N2 adsorption at 77k

F(B)	The (Gaussian) distribution function of the structural parameter B
F(L)	PSD of the heterogeneous solid adsorbent
F(Z)	Distribution function characterizing heterogeneity of microporous structure
GCMC	Grand canonical Monte Carlo simulation
h_{cr}	Critical capillary film thickness
H_{max}	Width of the largest pore
H_{min}	Width of the smallest pore
J	Local molar flux
J(X)	Micropore size distribution
L	Local pore size
m	Proportionality constant
NA	Avogadro number
q1–ql	Difference heat of condensation and adsorption
R	Universal gas constant
R	Radius of the nanotube
r/R	Ratio of distance to radius
r_{cr}	Critical capillary radius
SHN	Shahsavand-Niknam PSD calculating algorithm
S_{meso}	Total surface area of mesopores
S_{mi}	Total surface area of micropores
S_t	Slope of low pressure of α_s-plot
T	Absolute temperature
t-plot	T-curve method
V(A)	Chracteristric adsorption curve
V(r)	Interaction potential of the fluid molecule
Verlet	A type of Simulation algorithm
V_{meso}	Mesopore volume
V_{micro}	Micropore volume
V_t	Maximum volume adsorbed
V_t-v	Unoccupied pore volume
W^0_{mi}	Limiting volume of adsorption
x	Radius of the equivalent hemispherical meniscus

$X(A)$	Adsorption potential distribution function
$X^*(A)$	Non- normalized adsorption potential distribution function
x_0	The micropore half width at the maximum of the adsorption curve(mean of gaussian distribution)
$X_{mi}(A)$	Micropore Adsorption potential distribution function
y	Bulk mole fraction in binary systems
Z	Quantity associated with the micropore size
Γ	The Euler gamma function
ΔH	Differential enthalpy
ΔS	Differential entropy
ε_{FC}	Len–Jones interaction parameter
Θ	Surface coverage(IHK model)
$\theta = a/a0_{mi}$	Relative adsorption
$\theta_{mi}(p/p_0)$	Relative surface coverage of microporous reference adsorbent
$\theta_{mic}(z,A)$	The local isotherm in uniform micropres
$\theta_s(p/p_0)$	Relative surface coverage of nonporous reference adsorbent
σ_{FC}	Len–Jones interaction parameter
$\Phi(r)$	Extemal potential function
ω, Ω	Cross-sectional area
$\Omega[\rho(z)]$	Grand potential function
A	Adsorbate
$F[\rho(r)]$	The intrinsic Helmholtz free energy function
$J(L - d_a)$	The effective pore size distribution
L/d_A	Reduced pore width
$TVFM$	Theory of the volume filling of micropores
V	The pore volume
b	An equilibrium adsorption constant
C	Average interstitial concentration
f	Fluid fugacity
$f(H)$	Pore size distribution function
$f(z)$	Intrinsic molecular Helmholtz free energy of the adsorbate phase

k	Structural parameter
$m = (\beta\kappa)^{-2}$	A proportional coefficient
mic	Calculated only for the range of micropores
n	The best-fit parameter of the Dubinin–Astakhov
p	Equilibrium pressure of an adsorbate
p/p_0	Relative pressure
$r(p/p_0)$	The pore radius
$t(p/p_0)$	Statistical thickness of the film
Θ	The degree of micropore fillingdefined by the Dubinin–Astakhov equation
α	Thermal coefficient of the limiting adsorption
β	The similarity coefficient
γ	The surface tension
δ	The displacement of the surface
δ, Δ	Dispertion
ε	Adsorption characterized energy
ζ	Constant equal to $(1/\kappa^n)$
η	Slope of linear segment of the α_s-plot
$\theta(L, P)$	Local adsorption isotherm (kernel)
κ	The characteristic constant
λ	The regularization coefficient
μ	Chemical potential
υ	Liquid molar volume
ρ	Parameter of the gamma distribution function
$\rho(P, H)$	The mean density of N2 at pressure P in a pore of width H
$\rho(z)$	Local density of the adsorbed fluid
ζ	Proportionality constant
σ	Collision diameter
$\varphi ij (r)$	The pair potential between two atoms
$\chi(D_{in})$	The normalized differential PSD function

PREFACE

This volume covers a wide range of adsorption activities of porous carbon (PC), CNTs, and carbon nanostructures that have been employed so far for the removal of various pollutants from water, wastewater, and organic compounds. The low cost, high efficiency, simplicity and ease in the up-scaling of adsorption processes using PC make the adsorption technique attractive for the removal and recovery of organic compounds. The activated carbon modification process has also been of interest to overcome some of the limitations of the adsorbents. Due to a large specific surface area and small, hollow, and layered structures, CNTs and carbon nanostructures have been investigated as promising adsorbents for various metal ions; inorganic and organic pollutants can be easily modified by chemical treatment to increase their adsorption capacity.

There is the huge hope that nanotubes applications will lead to a cleaner and healthier environment. A brief summary of these modeling methods is presented in this volume. In addition, two important simulation methods, the Monte Carlo and Molecular Dynamic methods, are discussed in this volume. However, the presence of micro and mesopores is essential for many researchers aiming to control micro or mesoporosity. The present volume attempts to give a general view of the recent activities on the study of pore structure control, with the application of novel simulation and modeling methods and the necessity and importance of this control. This volume also provides a brief overview of the methodology and modeling beside simulation methods for characterization of nanoporous carbons by using adsorption isotherm parameters.

— **A. K. Haghi, PhD, Sabu Thomas, PhD, and
Moein MehdiPour MirMahaleh**

CHAPTER 1

BASIC CONCEPTS AND AN OVERVIEW

CONTENTS

1.1 INTRODUCTION

This chapter is focused on the pore size controlling in carbon-based nano adsorbent to apply simulation and modeling methods and describe the recent activities about it. Significant progress has been made in this process throughout the recent years, essential emphasis is put on the controlling and applications of both micro and mesoporosity carbons. Many novel methods are proposed such as catalytic activation, polymer blend, organic gel and template carbonization for the control of mesopores, and analyzing micropore distribution in activated carbons assuming an array of semi-infinite, rigid slits of distributed width whose walls are modeled as energetically uniform graphite. Various kinds of pores in solid materials are classified according to the origin of the pores and the structural factors of the pores are discussed as well as the methods for evaluation of the pore size distribution with molecular adsorption (molecular resolution porosimetry), small angle X-ray scattering, mercury porosimetry, nuclear magnetic resonance, and thermoporosirmetry. The main aim in the controlling of micropores is to produce molecular sieving carbon (MSC), which applied in special membranes or produce some carbon composites which remove special contaminate from aqueous environments as adsorbent or EDLCs in electrochemical application or aerogel carbon structures. Despite difference in particle size, the adsorption properties of activated carbon and carbon-based nano adsorbents (ACF-ACNF-CNF-CNT and CNT-Composites) are basically the same because the characteristics of activated carbon (pore size distribution, internal surface area and surface chemistry) controlling the equilibrium aspects of adsorption are independent of particle. The excellently regular structures of CNTs and its composites facilitate accurate simulation of CNT behavior by applying a variety of mathematical and numerical methods. The most important of this models include: Algorithms and simulation, such as Grand Canonical Monte Carlo (GCMC) simulation, Car-Parrinello molecular dynamics (CPMD), molecular dynamics simulation (MDS), LJ potential, DFT, HK, IHK, BJH, DA, DRS and BET Methods, ASA, Verlet and SHN algorithm and etc. Briefly, (in my PhD case study) this following research work include:

- Preparing web of CNT-Textile composite (carbon section in an appropriate furnace and other section by electros pining along CNT added in spinning dopp).
- Stabilization process performing in adequate temperatures in an appropriate furnace.
- Carbonizationprocess performing in adequate temperatures in an appropriate furnace.
- Activation process performing in adequate temperatures in an appropriate furnace.
- Adsorption isotherms and characterizing of this composite will be determined and fitted.

- Adsorption parameters like PSD, S_{BET}, V_{meso} and V_{mic} for CNT-Textile composite and Textile section (with more probability, ACNF), calculated by BET, DFT, DRS and HK models and results are compared.
- Pore size distributions and PSD curves are determined from experimental isotherms like t-plot.
- Samples simulated using GCMC or MDS by assumption is an adequate simulation cell (cylindrical slit-shape pore).
- Finally, experimental and model results of ACNF and CNT-ACNF and its effect on PSD and selective adsorption is compared, adapted and model is verified.

Final aim of this research as application view, is preparing novel carbon nano adsorbent composite which operate as selective adsorption to remove a wide range of special contaminate from aqueous environments. This detailed review presents more discussion about carbon nanostructures adsorbents, modeling and simulation methods in different sections.

1.1.1 ACTIVATED CARBON: PROPERTIES AND APPLICATION

Activated Carbon (AC) has been most effective adsorbent for the removal of a wide range of contaminates from aqueous or gaseous environment. It is a widely used adsorbent in the treatment of wastewaters due to its exceptionally high surface areas which range from 500 to 1500 $m^2\,g^{-1}$, well developed internal microporosity structure [1]. While the effectiveness of ACs to act as adsorbents for a wide range of pollutant materials is well noted and more research on AC modification are presented due to the need to enable ACs to develop affinity for special contaminants removal from wastewater. [2]. It is, therefore, essential to understand the various important factors that influence the adsorption capacity of AC due to their modification so that it can be tailored to their specific physical and chemical attributes to enhance their affinities pollutant materials. These factors include specific surface area, pore size distribution, pore volume and presence of surface functional groups. Generally, the adsorption capacity increases with specific surface area due to the availability of adsorption site while pore size, and micropore distribution are closely related to the composition of the AC, the type of raw material used and the degree of activation during production stage [3]. Here, we summarize the various AC modification techniques and their effects on adsorption of chemical species from aqueous solutions. Modifications of AC in granular or powdered form were reviewed. Based on extensive literature reviews, the authors have categorized the techniques into three broad groups, namely, modification of chemical, physical and biological characteristics which are further subdivided into their pertinent treatment techniques. Table 1.1 lists and compares the advantages and disadvantages of existing modification techniques with regards to technical aspects that are further elucidated in the following sections [1–4].

TABLE 1.1 Technical Advantages and Disadvantages of Existing Modification Techniques.

Modification	Treatment	Advantages	Disadvantages
Chemical characteristics	Acidic	Increases acidic functional groups on AC surface. Enhances chelation ability with metal species	May decrease BET surface area and pore volume, May give off undesired SO_2 (treatment with H_2SO_4) or NO_2 (treatment with HNO_3) gases, Has adverse effect on uptake of organics
	Basic	Enhances uptake of organics	May, in some cases, decrease the uptake of metal ions
	Impregnation of foreign material	Enhances in-built catalytic oxidation capability	May decrease BET surface area and pore volume
Physical characteristics	Heat	Increases BET surface area and pore volume	Decreases oxygen surface functional groups
Biological characteristics	Bioadsorption	Prolongs AC bed life by rapid oxidation of organics by bacteria before the material can occupy adsorption sites	Thick biofilm encapsulating AC may impede diffusion of adsorbate species

While these characteristics are reviewed separately as reflected by numerous AC modification research, it should be noted that there were also research with the direct intention of significantly modifying two or more characteristic and that the techniques reviewed are not intended to be exhaustive. The adsorption capacity depends on the accessibility of the organic molecules to the microporosity, which is dependent on their size [5]. Activated carbon can be used for removing taste and odor (T&O) compounds, synthetic organic chemicals (SOCs), and dissolved natural organic matter (DOM) from water. PAC typically has a diameter less than 0.15 mm, and can be applied at various locations in a treatment system GAC, with diameters ranging from 0.5 to 2.5 mm, is employed in fixed-bed adsorbents such as granular media filters or post filters. Each of these factors must be properly evaluated in determining the use of activated carbon in a practical application. The primary treatment objective of activated carbon adsorption in a particular water treatment plant determines the process design and operation; multiple objectives cannot, in most cases, be simultaneously optimized [5, 35].

It is critical in either case to understand and evaluate the adsorption interactions in the context of drinking water treatment systems. Activated carbons are prepared from different precursors and used in a wide range of industries. Their preparation, structure and applications were reviewed in different books and reviews. High BET surface area and light weights are the main advantages of activated carbons. Usually activated carbons have a wide range of pore sizes from micropores to macropores, which shows a marked contrast to the definite, pore size of zeolites [41, 46].

Previous researches showed that one of the most effective approaches for increasing mesopore volume of activated carbon is to catalyze the steam activation process by using transition metals or rare earth metal compounds, which can promisingly promote mesopore formation. The mechanism of mesoporous development is that the activation reaction takes place in the vicinity of metal/oxide particles, leading to the formation of mesopores by pitting holes into the carbon matrix. However, transition metals or rare earth metal compounds are still expensive and can seldom be reused on large scale, which limits industrial application of these processes. OH plays three roles in the preparation of the activated carbons. First of all, it guarantees that the carbonization of coal is in solid phase. Besides, KOH can react with coal to form micropores, which provide "activation path" for steam, which can expand the micropore. However, if there are too much micropores, combination of micropores will happen, resulting in the formation of much macropore, and then can increase the ignition loss, so there must be an optimal addition of KOH in the precursor. Furthermore, KOH catalyzes the steam activation process. Because of high catalytic activity of KOH, acid washing process is used after carbonization process and before steam activation process to vary the content of K-containing compounds left in the char, and then the degree of steam activation can probably be varied. Consequently, regulation of the pore size distribution of activated carbon from coal is performed in two ways: the first is to change the addition of KOH in the precursor to produce optimal micropore in the char, and the second is to use acid washing process to vary the degree of catalytic effect of K-containing compounds in steam activation. Modifying the steam activation method successfully regulated the pore size distributions of coal-based activated carbons. The principles for the regulation of pore size distribution in the activated carbons were also discussed in this article, and we found that regulation of the pore size distributions of the activated carbons from coal are performed in two ways: the first is to change the addition of KOH in the precursor to produce optimal micropore in the char, and the second is to use the acid washing process to vary the degree of catalytic effect of K-containing compounds in steam activation [35, 40].

1.1.2   ACTIVATED CARBON FIBERS (ACFS): PROPERTIES AND APPLICATION

Activated carbons fibers have been prepared recently and developed a new field of applications. They have a number of advantages over granular activated carbons. The principal merit to prepare activated carbon in fibrous morphology is its particular pore structure and a large physical surface area. Granular activated carbons have different sizes of pores (macropores, mesopores and micropores), whereas ACFs have mostly micropores on their surfaces. In granular activated carbons, adsorbates always have to reach micropores by passing through macropores and mesopores, whereas in ACFs they can directly reach most micropores because micropores are open to the outer surface and hence, exposed directly to adsorbates. Therefore, the adsorption rate, as well as the amount of adsorption, of gases into ACFs is much higher than those into granular activated carbons [1–3].

In recent work, the amount of adsorption of toluene molecules is much higher, and desorption proceeds faster in ACFs than granular activated carbons, effective elimination of SO• from exhausted gases by using ACFs was too. A very high specific surface area up to 2500 m^2 g^{-1} and a high micropore volume up to 1.6 cm^3 g^{-1} can be obtained in isotropic-pitch-based carbon fibers. For the preparation of these carbon fibers with a very high surface area such as 2500 m^2 g^{-1}, precursors which give a carbon with poor crystallinity are recommended; thus, mesophase-pitch-based carbon fibers did not give a high surface area, whereas isotropic pitch based carbon fibers did. Other advantage of ACFs is the possibility to prepare woven clothes and non-woven mats, which developed new applications in small purification systems for water treatment and also as a deodorant in refrigerators in houses, recently reported. In order to give the fibers an antibacterial function and to increase their deodorant function, some trials on supporting minute particles of different metals, such as Ag, Cu and Mn, were performed. Table 1.2 presents the-comparison between properties of activated carbon fibers and granular activated carbons [13–46].

TABLE 1.2   Comparison Between Some Properties of ACF and GAC.

	Activated Carbon Fibers	Granular Activated Carbons
S_{BET} ($m^2\,g^{-1}$)	700–2500	900–1200
Surface area ($m^2\,g^{-1}$)	0.2–2.0	~0.001
Mean diameter of pores (nm)	<40	Frommicro to macropores

Activated carbon fibers (ACFs), due to their microporosity, are an excellent material for a fundamental study of H$_2$adsorption capacity and enthalpy. Synthesized from polymeric carbon precursors, ACFs contain narrow and uniform pore size distributions with widths on the order of 1 nm. Images of ACFs from scanning tunneling microscopy have revealed networks of elongated slit-shaped and ellipsoid shaped pores. Edge terminations in graphitic layers are thought to be the most reactive sites during the steam/carbon dioxide activation process, resulting in a gradual lengthening of slit-shaped pores as a function of burn-off. ACFs subjected to less burn-off will have smaller pore volumes and a greater abundance of narrow pores widths. With longer activation times, the pore volume increases and the pores grow wider. This offers a convenient control for an experimental study of the correlation between pore structure and hydrogen adsorption. In the current study, the pore size distribution (PSD) of activated carbon fibers is used to interpret the enthalpy and the capacity of supercritical H_2 adsorption [2].

1.1.3 MOLECULAR SIEVING CARBONS (MSCS): PROPERTIES AND APPLICATION

Molecular sieving carbons (MSCs) have a smaller pore size with a sharper distribution in the range of micropores in comparison with other activated carbons for gas and liquid phase adsorbates. They have been used for adsorbing and eliminating pollutant samples with a very low concentration (ethylene gas adsorption to keep fruits and vegetables fresh, filtering of hazardous gases in power plants, etc.) An important application of these MSCs was developed in gas separation systems [1, 2]. The adsorption rate of gas molecules, such as nitrogen, oxygen, hydrogen and ethylene, depends strongly on the pore size of the MSC; the adsorption rate of a gas becomes slower for the MSC with the smaller pore size. The temperature also governs the rate of adsorption of a gas because of activated diffusion of adsorbate molecules in micropores: the higher the temperature, the faster the adsorption [47–49].

By controlling (swinging) these parameters, temperature and pressure of adsorbate gas, gas separation can be performed. Depending on which parameter is controlled, swing adsorption method is classified into two modes; temperature swing adsorption (TSA) and pressure swing adsorption (PSA). Adsorption of oxygen into the MSC completes within 5 min, but nitrogen is adsorbed very slowly, less than 10% of equilibrium adsorption even after 15 min. From the column of MSC, therefore, nitrogen rich gas comes out on the adsorption process, and oxygen-rich gas is obtained on the desorption process. By using more than two columns of MSC and repeating these adsorption/desorption processes, nitrogen gas is isolated from oxygen. This swing adsorption method for gas separation has advantages such as low energy cost, room temperature operation, and compact equipment [50–53].

1.1.4 SUPER-CAPACITORS (EDLCS) AS NEWEST APPLICATION OF CONTROLLING OF PORE SIZE IN POROUS CARBONS

The main objective of this paper is to provide a brief review of the pore size control that is an important factor influencing application of activated carbon or carbon nano structures adsorption in drinking water treatment or other adsorbent application. Different pore sizes in carbon materials are required for their applications. Therefore, the PSD in carbon materials has to be controlled during their preparation, by selecting the precursor, process and condition of carbonization, and also those of activation. A wide range broad distribution in pore size and shape is usually obtained in carbon materials. The control of pore size in carbons is essential in order to compete in adsorption performance with porous inorganic materials such as silica gels and zeolites, and to use the advantages of carbon materials such as high chemical stability, high temperature resistance and low weight. For applications in modern technology fields, not only high surface area and large pore volume but also a sharp pore size distribution at a definite size and control of surface nature of pore walls are strongly required. In order to control the pore structure in carbon materials, studies on the selection of precursors and preparation conditions have been extensively carried out and certain successes have been achieved [1–3]. Pore sizes and their distributions in adsorbents have to comply with requirements from different applications. Thus, relatively small pores are needed for gas adsorption and relatively large pores for liquid adsorption, and a very narrow PSD is required for molecular sieving applications. Macropores in carbon materials were found to be effective for sorption of viscous heavy oils. Recent novel techniques to control pore structure in carbon materials can be expected to contribute to overcome this limitation [41, 46].

One of newest application that shows importance of pore structure control in carbon materials, is an electric double-layer capacitor (EDLC) or super capacitor that is an energy storage device that uses the electric double layer formed at the interface between an electrode and the electrolyte. EDLCs are well documented to exhibit significantly higher specific powers and longer cycle lifetimes compared with those of most of rechargeable batteries, including lead acid, Ni-MH, and Li-ion batteries. Hence, EDLCs have attracted considerable interest, given the ever-increasing demands of electric vehicles, portable electronic devices, and power sources for memory backup. The capacitance of an EDLC depends on the surface area of the electrode materials. Therefore, activated carbons are necessary materials for EDLC electrodes because of their large surface area, highly porous structure, good adsorption properties, and high electrical conductivity. The electrochemical performance of EDLCs is related to the surface area, the pore structure, and the surface chemistry of the porous carbon. Various types of porous carbon have been widely studied for use as electrode materials for EDLCs. Their unusual structural and electronic properties make the carbon nanostructures applicable in, inter alia, the electrode materials of EDLCs and batteries. Activated carbon nanofibers are expected to be more useful than spherical activated carbon in allowing the relationship between pore structure and electrochemical properties to be investigated to prepare the polarizable electrodes for experimental EDLCs, EDLCs are well documented to exhibit significantly higher specific powers and longer cycle lifetimes compared with those of most of rechargeable batteries, including lead acid, Ni-MH, and Li-ion batteries [20, 34, 45].

Hence, EDLCs have attracted considerable interest, given the ever-increasing demands of electric vehicles, portable electronic devices, and power sources for memory backup. The capacitance of an EDLC depends on the surface area of the electrode materials. Therefore, activated carbons are necessary materials for EDLC electrodes because of their large surface area, highly porous structure, good adsorption properties, and high electrical conductivity. The electrochemical performance of EDLCs is related to the surface area, the pore structure, and the surface chemistry of the porous carbon. Various types of porous carbon have been widely studied for use as electrode materials for EDLCs. Their unusual structural and electronic properties make the carbon nanostructures applicable in the electrode materials of EDLCs and batteries. The principle of electrochemical capacitors, physical adsorption/desorption of electrolyte ions in solution, was applied for water purification by using different carbon materials [108, 113].

This work is concerned with such pore control methods proposed by researchers that their ultimate goal is to establish a method with tailoring carbon material pore structures to reach any kind of application. Researchers would much like to effort that have made to control micro and mesopores in carbon materials, and prepare them in achieving the final goal. The presence of mesopores in electrodes based on CNTs, due to the central canal and entanglement enables easy access of ions from electrolyte. For electrodes built from multi walled carbon nanotubes (MWCNTs), specific capacitance in a range of 4–135 F/g was found. For single walled carbon nanotubes (SWCNTs) a maximum specific capacitance of 180 F/g is reported. A comparative investigation of the specific capacitance achieved with CNTs and activated carbon material reveals the fact activated carbon material exhibited significantly higher capacitance. Super capacitor CNTs-based electrodes were fabricated by direct synthesis of nanotubes on the bulk Ni substrates, by means of plasma enhanced chemical vapor deposition of methane and hydrogen. The specific capacitance of electrodes with such nanotubes was of 49 F/g. MWCNTs were electrochemically oxidized and their performance in EDLCs was studied [45, 64, 68].

1.1.5 CARBON NANOTUBE (CNT): PROPERTIES AND APPLICATION

An article by Iijima that showed that carbon nanotubes are formed during arc-discharge synthesis of C_{60}, and other fullerenes also triggered an outburst of the interest in carbon nanofibers and nanotubes. These nanotubes may be even single walled, whereas low-temperature, catalytically grown tubes are multi-walled.

It has been realized that the fullerene-type materials and the carbon nanofibers known from catalysis are relatives and this broadens the scope of knowledge and of applications. This chapter describes the issues around application and production of carbon nanostructures. Electro spinning is a simple and versatile method for generating ultrathin fibers from a rich variety of materials that include polymers, nanocomposites and ceramics. In a typical process, an electrical potential is applied between a droplets of a polymer solution, or melt, held at the end of a capillary tube and a grounded target. When the applied electric field overcomes the surface tension of the droplet, a charged jet of polymer solution is ejected. The following parameters and processing variables affect the electro spinning process: (i) system parameters such as molecular weight, molecular weight distribution and architecture (branched, linear, etc.) of the polymer, and polymer solution properties (viscosity, conductivity, dielectric constant, and surface tension, charge carried by the spinning jet) and (ii) process parameters such as electric potential, flow rate and concentration, distance between the capillary and collection screen, ambient parameters (temperature, humidity and air velocity in the chamber) and finally motion of the target screen. Morphological changes can occur upon decreasing the distance between the syringe needle and the substrate. Increasing the distance or decreasing the electrical field decreases the bead density, regardless of the concentration of the polymer in the solution. Elemental carbon in the sp^2 hybridization can form a variety of amazing structures. The nanotubes consisted of up to several tens of graphitic shells (so-called multi-walled carbon nanotubes (MWNT)) with adjacent shell separation of 0.34 nm, diameters of 1 nm and high length/diameter ratio. Two years later, Iijima and Ichihashi synthesized single-walled carbon nanotubes (SWNT). There are two main types of carbon nanotubes that can have high structural perfection. Single walled nanotubes (SWNT) consist of a single graphite sheet seamlessly wrapped into a cylindrical tube. Multi-walled nanotubes (MWNT) comprise an array of such nanotubes that are concentrically nested like rings of a tree trunk [54, 57].

Recent discoveries of fullerene, a zero dimensional form of carbon and carbon nanotube, which is a one-dimensional form, have stimulated great interest in carbon materials overall. Fullerenes are geometric cage-like structures of carbon atoms that are composed of hexagonal and pentagonal faces when a Bucky ball is elongated to form a long and narrow tube of few nanometers diameter approximately, which is the basic form of carbon nanotube. This stimulated a frenzy of activities in properties measurements of doped fullerenes. The discovery of fullerenes led to the discovery of carbon nanotubes by Iijima in 1991. The discovery of carbon nanotubes created much excitement and stimulated extensive research into the properties of nanometer scale cylindrical carbon networks. Many researchers have reported mechanical properties of carbon nanotubes that exceed those of any previously existing materials. Although there are varying reports in the literature on the exact properties of carbon nanotubes, theoretical and experimental results have shown extremely high modulus, greater than 1 TPa (the elastic modulus of diamond is 1.2 TPa) and reported strengths 10–100 times higher than the strongest steel at a fraction of the weight. Indeed, if the reported mechanical properties are accurate, carbon nanotubes may result in an entire new class of advanced materials [58, 62].

In addition to the exceptional mechanical properties associated with carbon nanotubes, they also possess superior thermal and electric properties such as thermally stableup to 2800°C in vacuum, thermal conductivity about twice as high as diamond, electric current carrying capacity 1000 times higher than copper wires. These exceptional properties of carbon nanotubes have been investigated for devices such as field emission displays, scanning probe microscopy tips and microelectronic devices. Carbon nanotubes present significant opportunities to basic science and nanotechnology, and pose significant challenge for future work in this field. The approach of direct growth of nanowires into ordered structures on surfaces is a promising route to approach nanoscale problem and create novel molecular scale devices with advanced electrical, electromechanical and chemical functions [54].

1.1.5.1 STRUCTURE AND PROPERTIES

1.1.5.1.1 ELECTRICAL PROPERTIES

The Unique Electrical Properties of carbon nanotubes are to a large extent derived from their 1-D character and the peculiar electronic structure of graphite. They have extremely low electrical resistance. Resistance occurs when an electron collides with some defect in the crystal structure of the material through which it is passing. The defect could be an impurity atom, a defect in the crystal structure, or an atom vibrating. Such collisions deflect the electron from its path, but the electrons inside a carbon nanotube are not so easily scattered. Because of their very small diameter and huge ratio of length to diameter a ratio that can be up in the millions or even higher. In a 3-D conductor, electrons have plenty of opportunity to scatter, since they can do so at any angle. Any scattering gives rise to electrical resistance. In a 1-D conductor, however, electrons can travel only forward or backward. Under these circumstances, only backscattering (the change in electron motion from forward to backward) can lead to electrical resistance. But backscattering requires very strong collisions and is thus less likely to happen. So the electrons have fewer possibilities to scatter. This reduced scattering gives carbon nanotubes their very low resistance. In addition, they can carry the highest 30 current density of any known material, measured as high as 109 A/cm^2. One use for nanotubes that has already been developed is as extremely fine electron guns, which could be used as miniature cathode ray tubes (CRTs) in thin high-brightness low-energy low-weight displays. This type of display would consist of a group of many tiny CRTs, each providing the electrons to hit the phosphor of one pixel, instead of having one giant CRT whose electrons are aimed using electric and magnetic fields. These displays are known as Field Emission Displays (FEDs). A nanotube formed by joining nanotubes of two different diameters end to end can act as a diode, suggesting the possibility of constructing electronic computer circuits entirely out of nanotubes. Nanotubes have been shown to be superconducting at low temperatures [55].

1.1.5.1.2 MECHANICAL PROPERTIES

The carbon nanotubes are expected to have high stiffness and axial strength as a result of the carbon–carbon sp^2 bonding. The practical application of the nanotubes requires the study of the elastic response, the inelastic behavior and buckling, yield strength and fracture. Efforts have been applied to the experimental and theoretical investigation of these properties. Nanotubes are the stiffest known fiber, with a measured Young's modulus of 1.4 TPa. They have an expected elongation to failure of 20–30%, which combined with the stiffness, projects to a tensile strength well above 100 GPa (possibly higher), by far the highest known. For comparison, the Young's modulus of high-strength steel is around 200 GPa, and its tensile strength is 1–2 GPa [56].

1.1.5.1.3 THERMAL PROPERTIES

Prior to CNT, diamond was the best thermal conductor. CNT have now been shown to have a thermal conductivity at least twice that of diamond. CNT have the unique property of feeling cold to the touch, like metal, on the sides with the tube ends exposed, but similar to wood on the other sides. The specific heat and thermal conductivity of carbon nanotube systems are determined primarily by phonons. The measurements yield linear specific heat and thermal conductivity above1 K and below room temperature while a 0.62 behavior of the specific heat was observed below1 K. The linear temperature dependence can be explained with the linear k-vector dependence of the frequency of the longitudinal and twist acoustic phonons. The specific heat below1 K can be attributed to the transverse acoustic phonons with quadratic k dependence. The measurements of the thermoelectric power (TEP) of nanotube systems give direct information for the type of carriers and conductivity mechanisms [57].

1.1.5.2 STRUCTURE OF CARBON NANOTUBES

The multilayered nanotubes were found in the cathode tip deposits that form when a DC arc is sustained between the graphite electrodes of a fullerene generator. They are typically composed of 2 to 5 concentric cylindrical shells, with outer diameter typically a few tens of nanometer and lengths of the order of micrometer. Each shell has the structure of a rolled up graphene sheet with the sp^2 carbons forming a hexagonal lattice. The discovery of nanotubes has revolutionized researches in different directions. A light and high strength nanotube would be an ideal structural member for designing nanostructural instruments. It has reported that nanotubes could become as familiar as silicon in this century and the full development of the nanotubes would be around 2010. In order to familiarize the uses and applications of the nanotubes and their related products, an understanding of the structure, characterization and properties of the nanotubes is essential. The nanotubes possess conducting properties ranging from metallic to moderate band gap semiconductor. In general, the nanotubes could be specified in terms of the tube diameter (d) and the chiral angle (θ). The chiral vector (Ch) is defined as a line connected from two crystallo graphically equivalent sites on a two-dimensional graphene structure. The chiral vector can be defined in terms of the lattice translation indices (n, m) and the basic vectors a_1 and a_2 of the hexagonal lattice (a layer of grapheme sheet). The chiral angle (θ) is measured as an angle between the chiral vector Ch with respect to the zigzag direction (n, 0), where $\theta = 0°$ and the unit vectors of a_1 and a_2. The armchair nanotube is defined as the $\theta = 30°$ and the translation indices is (n, n). All other types of nanotubes could be identified as a pair of indices (n, m) where n $\neq$ m. The electronic conductivity is highly sensitive to a slight change of these parameters, which cause a changing of materials between metallic and semiconductor statues. Recently, it has been reported that the scanning tunneling microscopy (STM) and spectroscopy could be used to observe the electronic properties and atomic arrangement of SWNTs [55–57].

1.1.5.2.1   MULTI AND SINGLE WALL NANOTUBES

Multi-walled carbon nanotubes were first reported by Iijima in 1991 in carbon made by an arc-discharge method4. About two years later, he made the observation of single-walled nanotubes (SWNTs). A SWNT is a graphene sheet rolled over into a cylinder with typical diameter of the order of 1.4 nm.

Similar to that of a C_{60} Bucky ball, a MWNT consists of concentric cylinders with an interlayer spacing of 3.4Å and a diameter typically of the order of 10–20 nm. The lengths of the two types of tubes can be up to hundreds of microns or even centimeters. A SWNT is a molecular scale wire that has two key structural parameters. By folding a graphene sheet into a cylinder so that the beginning and end of a (m, n) lattice vector in the graphene plane join together, one obtains an (m, n) nanotube. The (m, n) indices determine the diameter of the nanotube, and also the so-called chirality [55].

1.1.5.3   SYNTHESIS METHODS OF CNT

1.1.5.3.1   SYNTHESIS OF CNT

The MWNT were first discovered in the soot of the arc-discharge method by Iijima. This method had been used long before in the production of carbon fibers and fullerenes. It took two more years for Iijima and Ichihashi to synthesize SWNT by use of metal catalysts in the arc-discharge method in 1993. Significant progress was achieved by laser ablation synthesis of bundles of aligned SWNT with small diameter distribution. Catalytic growth of nanotubes by the chemical vapor decomposition (CVD) method was used too [54].

ARC-DISCHARGE:

In 1991, Iijima reported the preparation of a new type of finite carbon structures consisting of needlelike tubes. The tubes were produced using an arc discharge evaporation method similar to that used for the fullerene synthesis. The carbon needles, ranging from 4 to 30 nm in diameter and up to 1 mm in length, were grown on the negative end of the carbon electrode used for the direct current (dc) arc-discharge evaporation of carbon in an argon filled vessel (100 Torr). Iijima used an arc-discharge chamber filled with a gas mixture of 10 Torr methane and 40 Torr argon. Two vertical thin electrodes were installed in the center of the chamber. The lower electrode, the cathode, had a shallow dip to hold a small piece of iron during the evaporation. The arc-discharge was generated by running a dc current of 200 A at 20 V. Laser beam vaporizes target of a mixture of graphite and metal catalyst (Co, Ni) in a horizontal tube in a flow of inert gas at controlled pressure and in a tube furnace at 1200°C. The nanotubes are deposited on a water-cooled collector outside the furnace electrodes. The use of the three components argon, iron and methane, was critical for the synthesis of SWNT. The nanotubes had diameters of 1 nm with a broad diameter distribution between 0.7 and 1.65 nm. In the arc-discharge synthesis of nanotubes, used as anodes thin electrodes with bored holes, which were filled with a mixture of pure powdered metals (Fe, Ni or Co) and graphite. The electrodes were vaporized with a current of 95–105 A in 100–500 Torr of He. Large quantities of SWNT were generated by the arc-technique. The arc was generated between two graphite electrodes in a reactor under helium atmosphere (660 mbar)[56].

LASER-ABLATION:

In 1996, Smalley and co-workers produced high yields (>70%) of SWNT by laser ablation (vaporization) of graphite rods with small amounts of Ni and Co at 1200°C. The tube grows until too many catalyst atoms aggregate on the end of the nanotube. The large particles either detach or become over coated with sufficient carbon to poison the catalysis. This allows the tube to terminate with a fullerene like tip or with a catalyst particle. Both arc-discharge and laser-ablation techniques have the advantage of high (>70%) yields of SWNT and the drawback that (1) they rely on evaporation of carbon atoms from solid targets at temperatures >3000°C, and (2) the nanotubes are tangled which makes difficult the purification and application of the samples [56].

CHEMICAL VAPOR DEPOSITION (CVD):

Despite the described progress of synthetic techniques for nanotubes, there still remained two major problems in their synthesis, that is, large scale production and ordered synthesis. But in 1996, a CVD method emerged as a new candidate for nanotube synthesis. This method is capable of controlling growth direction on a substrate and synthesizing13 large quantity of nanotubes. In this process a mixture of hydrocarbon gas, acetylene, methane or ethylene and nitrogen was introduced into the reaction chamber. During the reaction, nanotubes were formed on the substrate by the decomposition of the hydrocarbon at temperatures 700–900°C and atmospheric pressure. The process has two main advantages: the nanotubes are obtained at much lower CVD reactor temperature, although this is at the cost of lower quality, and the catalyst can be deposited on a substrate, which allows for the formation of novel structures [56].

THE SUBSTRATE:
The preparation of the substrate and the use of the catalyst deserve special attention, because they determine the structure of the tubes. The substrate is usually silicon, but also, glass and alumina are used. The catalysts are metal nanoparticles, like Fe, Co and Ni, which can be deposited on silicon substrates either from solution, electron beam evaporation or by physical sputtering. The nanotube diameter depends on the catalyst particle size, therefore, the catalyst deposition technique, in particular the ability to control the particle size, is critical to develop nano devices. Porous silicon is an ideal substrate for growing self-oriented nanotubes on large surfaces. It has been proven that nanotubes grow at a higher ratio (length per minute), and they are better aligned than on plain silicon. The nanotubes grow parallel to each other and perpendicular to the substrate surface, because of catalyst surface interaction and the van der Waals forces developed between the tubes [56].

THE SOL–GEL:
The sol–gel method uses a dried silicon gel, which has undergone several chemical processes, to grow highly aligned nanotubes. The substrate can be reused after depositing new catalyst particles on the surface. The length of the nanotube arrays increases with the growth time, and reaches about 2 mm after 48h growth [56].

GAS PHASE METAL CATALYST:

In the methods described above, the metal catalysts are deposited or embedded on the substrate before the deposition of the carbon begins. A new method is to use a gas phase for introducing the catalyst, in whichboth the catalyst and the hydrocarbon gas are fed into a furnace, followed by catalytic reaction in the gas phase. The latter method is suitable for large-scale synthesis, because the nanotubes are free from catalytic supports and the reaction can be operated continuously. A high-pressure carbon monoxide (CO) reaction method, in which CO gas reacts with iron pentacarbonyl, $Fe(CO)_5$ to form SWNT, has been developed. SWNT have also been synthesized from a mixture of benzene and ferrocene, $Fe(C_5H_5)_2$ in a hydrogen gas flow. In both methods, catalyst nanoparticles are formed through thermal decomposition of organo metallic compounds, such as iron pentacarbonyl and ferrocene. The reverse micelle method is promising, which contains catalyst nanoparticles (Mo and Co) with a relatively homogeneous size distribution in a solution. The presence of surfactant makes the nanoparticles soluble in an organic solvent, such as toluene and benzene. The colloidal solution can be sprayed into a furnace, at a temperature of 1200°C; it vaporizes simultaneously with the injection and a reaction occurs to form a carbon product. The toluene vapor and metal nanoparticles act as carbon source and catalyst, respectively. The carbon product is removed from the hot zone of the furnace by a gas stream (hydrogen) and collected at the bottom of the chamber [55–56].

1.1.5.3.2   RECENT TRENDS IN THE SYNTHESIS OF CNT

Some researchers synthesized carbon nanotubes from an aerosol precursor. Solutions of transition metal cluster compounds were atomized by electro hydrodynamic means and the resultant aerosol was reacted with ethane in the gas phase to catalyze the formation of carbon nanotubes. The use of an aerosol of iron penta carbonyl resulted in the formation of multi-walled nanotubes, mostly 6–9 nm in diameter, whereas the use of iron dodecacarbonyl gave results that were concentration dependent. High concentrations resulted in a wide diameter range (30–200 nm) whereas lower concentrations gave multi-walled nanotubes with diameters of 19–23 nm. CNT synthesized by electrically arcing carbon rods in helium (99.99%) in a stainless steel chamber with an inner diameter of 600 mm and a height of 350 mm. The anode was a coal-derived carbon rod (10 mm in diameter, 100–200 mm in length); the cathode was a high-purity graphite electrode (16 mm in diameter, 30 mm in length). The helium gas functioned as buffer gas and its pressure was varied in range of 0.033–0.076 MPa in the experiment. CNT synthesized via a novel route using an iron catalyst at the extremely low temperature of 180°C. The carbon clusters can grow into nanotubes in the presence of Fe catalyst, which was obtained by the decomposition of iron carbonyl Fe$_2$(CO)$_9$ at 250°C under nitrogen atmosphere. SWNT have been successfully synthesized using a fluidized bed method that involves fluidization of a catalyst/support at high temperatures by a hydrocarbon flow. A new method, which combines non-equilibrium plasma reaction with template controlled growth technology, has been developed for synthesizing aligned carbon nanotubes at atmospheric pressure and low temperature. Multiwall carbon nanotubes with diameters of approximately 40 nm were restrictedly synthesized in the channels of anodic aluminum oxide template from a methane hydrogen mixture gas by discharge plasma reaction at a temperature below 200°C. In a recent technique, Nebulized spray pyrolysis, large-scale synthesis of MWNT and aligned MWNT bundles is reported. Nebulized spray is a spray generated by an ultrasonic atomizer. A SEM image of aligned MWNT bundles obtained by the pyrolysis of a nebulized spray of ferrocene–toluene–acetylene mixture. The advantage of using a nebulized spray is the ease of scaling into an industrial scale process, as the reactants are fed into the furnace continuously [55, 56].

During nanotube synthesis, impurities in the form of catalyst particles, amorphous carbon and non-tubular fullerenes are also produced. Thus, subsequent purification steps are needed to separate the carbon nanotubes. The gas phase processes tend to produce nanotubes with fewer impurities and are more amenable to large scale processing. It is believed that the gas phase techniques, such as CVD, for nanotube growth offer greater potential for the scaling up of nanotubes production for applications. Initially, electric-arc discharge technique was the most popular technique to prepare the SWNTs as well as MWNTs. In this technique, the carbon arc provides a simple and traditional tool for generating the high temperatures needed for the vaporization of carbon atoms into a plasma (>3000°C). The gas phase growth of single walled nanotubes by using carbon monoxide as the carbon source has already been reported. They found the highest yields of single walled nanotubes occurred at the highest accessible temperature and pressure (1200°C, 10 atm). They have modified this process to produce large quantities of single walled nanotubes with remarkable purity. The lower processing temperatures also enable the growth of carbon nanotubes on a wide variety of substrates. CVD method has been successful to produce the CNTs in large quantity, and also to obtain the vertically aligned CNTs at relatively low temperature. In particular, growth of vertically aligned CNTs on large substrate area at low temperature, for instance, softening temperature of the glass is an important factor for the practical application of the electron emitters to the field emission displays. A lot of reports on the synthesis of single walled as well as multi-walled carbon nanotubes using the plasma enhanced chemical vapor deposition and microwave plasma enhanced chemical vapor deposition techniques, are available in the literature. Others successfully synthesized vertically aligned carbon nanotubes (CNTs) at 550°C on Ni-coated Si substrate placed parallel to Pd plate as a dual catalyst and a tungsten wire filament. The bamboo shaped carbon nanotubes can be obtained only if the reaction temperature is higher than 1000 K, and carbon fibers can be obtained at lower temperatures. They have also discussed the role and state of the catalyst particles. They have found that a plug-shaped Ni particle always plunged at the top end of a nanotube. Experimental results indicated that the catalytic particles exist in a liquid state during the synthesis procedure. After having crystallized, the orientations of plunged Ni particles randomly distributed around the axis. All three metals deposited on the quartz plates are found to be efficient catalysts for the growth of CNTs in good alignment by CVD using ethylene diamine [56–57].

The CNTs are multi-walled with a bamboo-like graphitic structure. The hollow compartments of the tubes are separated by the graphitic interlink layers. This is a simple and efficient way for the production of carbon nanotubes with good order using ethylene diamine by CVD without plasma aid. Others produced CNTs by hotwire CVD using a mixture of C_2H_2 and NH_3 gases. They used crystalline Si and SiO_2 substrates coated with Ni films of 30 nm in thickness as a catalyst for nanotube formation. SEM images showed that micron sized grains were present in the deposit on the Si substrate, whereas nanosized grains were evident in the deposit on the SiO_2 substrate. Nanotube formation could not be confirmed by SEM, but there was evidence of the possible formation of nanotubes on the Ni-coated SiO_2 substrate. Two novel nanostructures, which are probably the missing link between onion like carbon particles and nanotubes have also been obtained. Low synthesis temperature <520°C due to the non-equilibrium characteristics of microwave plasma operated at low pressure is also reported, which is crucial for some fascinating applications. Some researchers employed a high-density plasma chemical vapor deposition (PECVD) to grow high quality carbon nanotubes at low temperatures. High density, aligned CNTs can be grown on Si and glass substrate. The CNTs were selectively deposited on the patterned Ni catalyst layer, which was sputtered on Si substrate. Some others synthesized pure carbon nanotubes at very low temperature using MW-PECVD with methane/hydrogen gas. Others prepared the massive carbon nanotubes on silicon, quartz and ceramic substrates using MW-PECVD. The nanotubes, ranging from 10 to 120 nm in diameter and a few tens of micron in length, were formed under hydrocarbon plasma at 720°C with the aid of iron-oxide particles. Morphology of the nanotubes is strongly influenced by the flow ratio of methane to hydrogen. Defectless nanotubes with small diameters are favorably produced under a small flow ratio. To date, many methods for synthesizing carbon nanotubes have been developed and most of which operated at high temperature over 4000°Cfor graphite arc discharge and laser ablation. These temperatures are too high and unsuitable for the fabrication of electronic devices because most electric connections are made of aluminum with the melting point below 700°C. These challenges have promoted the current exploration of low temperature synthesis of carbon nanotubes such as CVD, PECVD and MWPECVD. Recently, these low temperature methods have been successful in growing highly aligned and large quantities of carbon nanotubes. It is also possible to control over length, diameter and structure of carbon nanotubes grown by CVD techniques. Therefore, CVD techniques are the most popular methods to synthesize the carbon nanotubes [56].

Recently have grown straight carbon nanotubes, carbon nanotubes "knees," Y-branches of carbon nanotubes and coiled carbon nanotubes on a graphite substrate held at room temperature by the decomposition of fullerene under moderate heating at 450°C in the presence of 200-nm Ni particles. The grown structures were investigated without any further manipulation by STM. The formation of the carbon nanostructures containing non-hexagonal rings is attributed partly to the templating effect of the high pyrolitic graphite (HOPG), partly to the growth at room temperature, which enhances the probability of quenching-in for non-hexagonal rings. Similar coiled nanotubes were found after several steps of chemical treatment in a catalytically grown carbon nanotube sample. They have examined growth of CNTs on a porous alumina template in order to improve the selectivity and uniformity of CNTs. They have fabricated well ordered, nanosized pores on a Si substrate using an anodic oxidation method. Recently developed a simple process for selective removal of carbon from single-walled carbon nanotube samples based on a mild oxidation by carbon dioxide. Some nanotubes were found to be partially filled with a solid material (probably metallic iron) that seems to catalyze the nanotube growth. Some regions of the deposit also revealed the presence of nanoparticles. The present experimental conditions should be suitable to produce locally structured deposits of carbon nanotubes for various applications. Recently synthesized straight and bamboo like carbon nanotubes in a methane diffusion flame using a Ni–Cr–Fe wire as a substrate. The catalyst particles were nickel and iron oxides formed on the wire surface inside the flame. The carbon growth over the catalysts has been followed gravimetrically in situ. The reaction was stopped after different pulse numbers in an attempt to control both the diameters and the lengths of the carbon nanotubes, which were characterized by transmission electron microscopy (TEM). Recently prepared two kinds of catalytic layers onto n-typed silicon substrate nickel by sputtering and iron (III) nitrate metal oxide by spin coating. For iron (III) nitrate metal oxide 0.5 mol of ferric nitrate non-anhydrate($Fe_2(NO_3)_2$.$9H_2O$] ethanol solution was coated onto tubes on both Ni and iron (III) nitrate metal oxide layers by the HFPECVD (hot filament plasma enhanced chemical vapor deposition) method [55, 58].

1.1.5.4 GRAFTING OF POLYMERS

The covalent reaction of CNT with polymers is important because the long polymer chains help to dissolve the tubes into a wide range of solvents even at a low degree of functionalization. There are two main methodologies for the covalent attachment of polymeric substances to the surface of nanotubes, which are defined as "grafting to" and "grafting from" methods. The former relies on the synthesis of a polymer with a specific molecular weight followed by end group transformation. Subsequently, this polymer chain is attached to the graphitic surface of CNT. The "grafting from" method is based on the covalent immobilization of the polymer precursors on the surface of the nanotubes and subsequent propagation of the polymerization in the presence of monomeric species.

1.1.5.4.1 "GRAFTING TO" METHOD

Recently reported the chemical reaction of CNT and PMMA using ultrasonication. The polymer attachment was monitored by FT-IR and TEM. As a result of this grafting, CNT were purified by filtration from carbonaceous impurities and metal particles. A nucleophilic reaction of polymeric An alternative approach was reported by the group of researchers MWNT were functionalized with n-butyllithium and subsequently coupled with halogenated polymers. Microscopy images showed polymer-coated tubes while the blend of the modified material and the polymer matrix exhibited enhanced properties in tensile testing experiments. They reported the grafting of functionalized polystyrene to CNT via a cyclo addition reaction. An azido polystyrene with a defined molecular weight was synthesized by atom transfer radical polymerization and then added to nanotubes. In a different approach, chemically modified CNT with appended double bonds were functionalized with living polystyryllithium anions via anionicpolymerization. The resulting composites were soluble in common organic solvents. Using an alternative method, polymers prepared by nitroxide-mediated free radical polymerization were used to functionalize SWNT through a radical coupling reaction of polymer-centered radicals. The in situ generation of polymer radical species takes place via thermal loss of the nitroxide-capping agent. The polymer-grafted tubes were fully characterized by UV-vis, NMR, and Raman spectroscopies.

1.1.5.4.2   "GRAFTING FROM" METHOD

CNT-polymer composites were first fabricated by an in situ radical polymerization process. Following this procedure, the double bonds of the nanotube surface were opened by initiator molecules and the CNT surface played the role of grafting agent. Similar results were obtained by several research groups. Depending on the type of monomer, it was possible not only to solubilize CNT but also to purify the raw material from catalyst or amorphous carbon. Through the negative charges of the polymer chain, the composite could be dispersed in aqueous media, whereas the impurities were eliminated by centrifugation. In a subsequent work, the same authors fabricated films consisting of alternating layers of anionic PSS-grafted nanotubes and cationic diazopolymer. The ionic bonds in the film were converted to covalent bonds upon UV irradiation, which improved greatly the stability of the composite material. They prepared polyvinylpyridine (PVP)-grafted SWNT by in situ polymerization. Solutions of such composites remained stable for at least 8 months. Layer by layer deposition of alternating thin films of SWNT-PVP and poly(acrylic acid) resulted in freestanding membranes, held together strongly by hydrogen bonding. Assemblies of PSS-grafted CNT with positively charged porphyrins were prepared via electrostatic interactions. The nano assembly gave rise to photo induced intra complex charge separation that lives for tens of microseconds. The authors have demonstrated that the incorporation of CNT-porphyrin hybrids onto indium tin oxide (ITO) electrodes leads to solar energy conversion devices. The raw material was treated with sec-butyllithium, which introduces a carbanionic species on the graphitic surface and causes exfoliation of the bundles. When a monomer was added, the nanotube carbon ionsinitiate polymerization, resulting in covalent grafting of the polystyrene chains. They studied the fabrication of composites by insitu ultrasonic induced emulsion polymerization of acrylates. It was not necessary to use any initiating species, and the polymer chains were covalently attached to the nanotube surface. MWNT grafted with poly(methyl methacrylate) were synthesized by emulsion polymerization of the monomer in the presence of a radical initiator or a cross-linking agent. CNT were found to react mostly with radical-type oligomers. The modified tubes had an enhanced adhesion to the polymer matrix, as could be observed by the improved mechanical properties of the composite. A different approach to composite preparation involves the attachment of atom transfer radical polymerization (ATRP) initiators to the graphitic network. These initiators were found to be active in the polymerization of various acrylate monomers. Recently prepared and characterized composites of nanotubes with methyl methacrylate and tert-butyl acrylate. The former composites were found to be insoluble in common solvents, while the latter were soluble in a variety of organic media. The fabrication of nanotube polyaniline composites via in situ chemical polymerization of aniline was studied by many groups. Initially, a charge transfer interaction was suggested, whereas a covalent attachment between the two components was described. The surface modification of SWNT was reported recently via in situ Ziegler-Natta polymerization of ethylene. The exact mechanism of nanotube-polymer interaction remains unclear, althoughthe authors suggested that a possible cross-linking could take place between the two components. The development of an integrated nanotube-epoxy polymer composite was reported by some authors. In the fabrication process, the authors used functionalized tubes with amino groups at the ends. These moieties could react easily with the epoxy groups and act as curing agents for the epoxy matrix. The cross-linked structure was most likely formed through covalent bonds between the tubes and the epoxy polymer. Multi-walled CNT were successfully modified with polyacrylonitrile chains by applying electrochemical polymerization of the monomer. The surface-functionalized tubes showed a good degree of dispersion in DMF while further proofs of debundling were obtained by TEM images [56].

1.1.6   CNT-COMPOSITES BASED ADSORBENTS

Potential practical applications of CNTs such as chemical sensors, field emission, electronic devices, high sensitivity nanobalance for nanoscopic particles, nanotweezers, reinforcements in high performance composites, biomedical and chemical investigations, anode for lithium ion batteries, super capacitors and hydrogen storage have been reported. Even though the challenges in fabrication may prohibit realization of many of these practical device applications, the fact that the properties of CNTs can be altered by suitable surface modifications can be exploited for more imminent realization of practical devices. In this respect, a combination of CNTs and other nanomaterials, such as nanocrystalline metal oxide/CNTs, polymer/CNTs and metal filled CNTs may have unique properties and research have therefore been focused on the processing of these CNT based nano composites and their different applications [54, 57].

Adsorption of single metal ions, dyes and organic pollutants on CNT-based adsorbent composites is one of the most applications of this material. MWCNT/ iron oxide magnetic composites were prepared and used for adsorptions of several metal ions. The CNTs were purified by using nitric acid, which results in modification of the surface of the nanotubes with oxygen containing groups like carbonyl and hydroxyl groups. The adsorption capability of the composite is higher than that of nanotubes and activated carbon. The sorption of ions such as Pb(II) and Cu(II) ions on the composite were spontaneous and endothermic processes based on the thermo dynamic parameters (ΔH, ΔS and ΔG) calculated from temperature dependent sorption isotherms. Alumina-coated MWCNTs were synthesized and reported for its utilization as adsorbent for the removal of lead ions from aqueous solutions in two modes. With an increase in influent pH between 3 and 7, the percentage of lead removed increases. The adsorption capacity increases by increasing agitation speed, contact time and adsorbent dosage. The reported composite can be regenerated as it was confirmed by SEM and EDX analysis [58, 60].

Grafting of polymers to nanotubes has been realized via both"grafting-to" and "grafting-from" approaches that mention before. The 'grafting-to' method is based on attachment of premade end-functionalized polymer onto the tips and convex walls of the nanotubes via chemical reactions such as etherification and amidization. One advantage of this method is that the mass and distribution of the grafted polymers can be more precisely controlled. However, initial grafted polymer chains sterically prevent diffusion of additional polymer chains to the nanotube surface resulting in low grafting density. The 'grafting-from' method is based on immobilization of reactive groups (initiators) onto the surface, followed by in situ polymerization of appropriate monomers to form polymer-grafted nanotubes. The advantage of this approach is that very high grafting density can be achieved. But careful control of the amount of initiator and the conditions for the polymerization reaction is required. Rather than the 'grafting-to' method used by some researchers. They investigated the 'grafting-from' approach, which involves the propagation of dendrimers from CNT surfaces by in situ polymerization of monomers in the presence of CNT attached macro initiators [73].

The mobility and small size of the fugitive monomers, in contrast to that of preformed dendritic polymers, are expected to improve the efficiency and yield of the 'grafting-from' approach and the resulting debundling process. The number of peripheral reactive groups can be precisely controlled by choosing the appropriate synthetic generation. Functionalization of nanotubes with dendrimers represents a particularly promising strategy to attach an in principle unlimited number of functional groups onto the SWNT surfaces, and thus to significantly increase the compatibility and reactivity of CNTs with thermosetting polymer matrices, such as epoxy, bismaleimide, and cyanate ester. Finally, to achieve good alignment of nanotubes to exploit their superior anisotropic mechanical properties, they applied the reactive spinning process. Spinning has been widely used for thermoplastic resins such as polyvinyl acetate (PVA), polyamide (PA), and polycarbonate (PC), but it has rarely been applied to thermosetting resins, except for cyanate ester. Control of viscosity within a narrow range suitable for spinning is often tricky and resin-specific. By comparison with cyanate ester, thermosetting epoxy generally has a smaller range of spinning conditions due to its higher reactivity. However, use of thermo set resins as the matrix for fiber spinning offers advantages of lower viscosity and reaction with functional groups on the nanotubes [72].

Although, several reports demonstrated that CNTs have good adsorption capacities for dyes due to their hollow and layered nanosized structures, which in turn have a large specific surface area, but their high cost restricts their use from industrial application. Furthermore, separation of CNTs from aqueous solution is very difficult because of its smaller size and high aggregation property. The problem of cost and separation of CNTs from aqueous medium can be overcome by making composites of CNTs with polymers, metal oxide, carbon etc. Such materials act as a stable matrix to the CNTs. Recently the CNTs have been used as a promising nanofiller for the preparation of CNT based nanocomposites because of their excellent improved adsorption, mechanical, electrical and thermal properties. CNT based composites are expected to be excellent adsorbent because CNTs provide not only the additional active sites but also larger surface area, which in turn makes them good adsorbents compared to their parent materials [72].

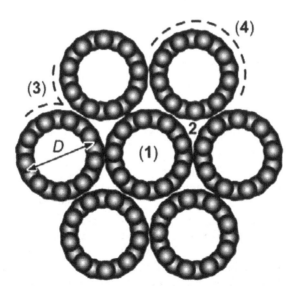

FIGURE 1.1 Different adsorption sites in a homogeneous bundle of SWCNTs with tube diameter D: (1) intratubular, (2) interstitial channel, (3) external groove, and (4) exposed surface of peripheral tube. Sites 1 and 2 comprise the internal porous volume of the bundle, whereas sites 3 and 4 are both located on the external surface of the bundle.

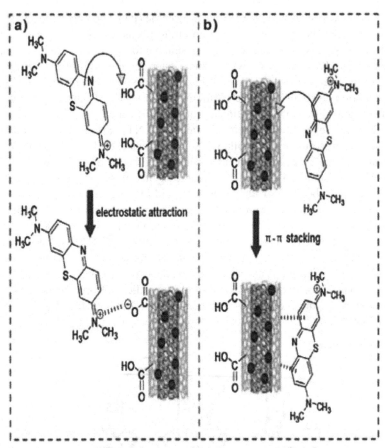

FIGURE 1.2   Schematic illustration of the possible interaction between MWCNTs and methylene blue: (a) electrostatic attraction and (b) π–π stacking.

CNTs can be combined with various metal oxides for the degradation of some organic pollutants too. Carbon nanotubes/metal oxide (CNT/MO) composites can be prepared by various methods such as wet chemical, sol gel, physical and mechanical methods. To form nanocomposite, CNTs can be combined with various metal oxides like Ti_2O_3, ZnO, WO_3, Fe_2O_3, and Al_2O_3. The produced nanocomposite can be used for the removal of various pollutants. Nanoscale Pd/Fe particles were combined with MWNTs and the resulted composite was used to remove2,4-dichlorophenol (2,4-DCP). It was reported that the MB adsorption was pH-dependent and adsorption kinetics was best described by the pseudo second-order model. Iron oxide/CNT composite was reported to be efficient adsorbent for remediation of chlorinated hydrocarbons. The efficiency of some other nanocomposites like CNT/alumina, CNT/titania and CNT/ZnO has also been reported [60–62].

Chitosan (CS) is one of the best adsorbents for the removal of dyes due to its multiple functional groups, biocompatibility and biodegradability, but its low mechanical strength limits its commercial applications. Impregnation of CS hydrogel beads with CNTs (CS/CNT beads) resulted in significant improved mechanical strength. In CS/CNT composite, CNTs and CS are like a symbiosis, CNTs help to improve the mechanical strength of CS while CS help to reduce the cost of CNTs for adsorption, while the resulted composite solves the problem of separating CNTs from aqueous medium. To resolve the aggregation and dispersion problem of CNTs, prepared the CNTs/activated carbon fabric (CNTs/ACF) composite and its application was investigated for the removal of phenol and basic violet 10 (BV10). CNTs/ACF was prepared via directly growing nanoscaled CNTs on microscaled carbon matrix. Polyacrylo nitrile was used as a source of carbon. From the results, it was observed that dye adsorption equilibrium time for CNTs/ACF is shorter as compared to ACF and monolayer adsorption capacity does not display a linear increase with increasing the BET surface area. The decoration of CNTs tends to lower the porosity of the ACF from 1065 to 565 m^2/g. This finding indicates that the total microporosity of ACF cannot be fully accessed by the dye molecules. Therefore, the appearance of CNTs plays a positive role in (i) facilitating the pore accessibility to adsorbates and (ii) providing more adsorptive sites for the liquid-phase adsorption. This reflects that CNTs/ACF contains a large number of mesopore channels, thus preventing the pore blockage from the diffusion path of micropores for adsorbates to penetrate [72, 73].

Incorporation of magnetic property in CNTs is another good technique to separate CNTs from solution. The magnetic adsorbent can be well dispersed in the water and easily separated magnetically. Magnetic-modified MWCNTs were used for removal of cationic dye crystal violet (CV), thionine (Th), janus green B (JG), and methylene blue (MB). To find the optimum adsorption, effect of various parameters including initial pH, dosages of adsorbent and contact time have been investigated. The optimum adsorption was found to be at pH 7.0 for all dyes. The removal efficiency of cationic dyes using GG/MWCNT/Fe_3O_4 is higher as compared with other adsorbents such as MWCNTs and MWCNT/Fe_3O_4. The magnetic GG/MWCNT/Fe_3O_4 possesses the high adsorption properties and magnetic separation and can therefore be used as magnetic adsorbents to remove the contaminants from aqueous solutions. A novel magnetic composite bio adsorbent composed of chitosan wrapping magnetic nanosized γ-Fe_2O_3 and multi-walled carbon nanotubes (m-CS/γ-Fe_2O_3/MWCNTs) was prepared for the removal of methyl orange. The adsorption capacity of MO onto m-CS/γ-Fe_2O_3/MWCNTs was 2.2 times higher than m-CS/γ-Fe_2O_3. The adsorption capacity of MO onto m-CS/γ-Fe_2O_3/MWCNTs was also higher than MWCNTs. Kinetics data and adsorption isotherm data were better fitted by pseudo second-order kinetic model and by Langmuir isotherm, respectively [58, 62, 72, 73].

Recently developed a novel adsorbent by inserting MWCNTs into the cavities of dolomite for scavenging of ethidium the foam line CNTs/dolomite adsorbent. Foam like ternary composite PUF/diatomite/dispersed-MWCNTs, gave the highest capacities for adsorption of these dyes, followed by PUF/agglomerated-MW-CNTs, and then PUF/dispersed-MWCNTs. Adsorption isotherm study revealed the monolayer adsorption at higher concentration and multilayer adsorption at lower concentration. Pseudo first order kinetics gives the best-fitted results compared to the pseudo second order. Self-assembled cylindrical graphene–MCNT (G–MCNT) hybrid, synthesized by the one pot hydrothermal process was used as adsorbent for the removal of methylene blue in batch process. G–MCNT hybrid showed good performance for the removal [72, 73].

Most important CNT-based as adsorbent composites include: CNT–Chitosan, CNT–ACF, CNT–Dolomite, CNT–Cellulose, CNT–Magnetic and Metal Oxide composites, CNT–Fiber composites, CNT–CF(PAN),CNT–Alginate, CNT–PANI composites and CNTs–Graphene. These composites are investigated in next section.

1.1.6.1 CNT-CHITOSAN COMPOSITES AS ADSORBENT

Chitosan (CS) is one of the best adsorbents for the removal of dyes due to its multiple functional groups, biocompatibility and biodegradability, but its low mechanical strength limits its commercial applications. Impregnation of CS hydrogel beads with CNTs (CS/CNT beads) resulted in significant improved mechanical strength. In CS/CNT composite, CNTs and CS are like a symbiosis, CNTs help to improve the mechanical strength of CS while CS help to reduce the cost of CNTs for adsorption, while the resulted composite solves the problem of separating CNTs from aqueous medium. Impregnation of 0.05 wt.% cetyl trimethyl ammonium bromide (CTAB) increased the maximum adsorption capacity of CS beads from media. The small difference in maximum adsorption capacity of CS/CNT beads and CS/CTAB beads indicated that CTAB molecules played a significant role in enhancing the adsorption performance of both varieties of beads. However, higher maximum adsorption capacity of CS/CNT beads than CS/CTAB beads suggested that CNTs itself in the beads adsorb CR during adsorption. This work suggests that surfactant played an important role in the removal of dyes. In his further studies [72, 73] on CS/CNTs, some researchers reported the effect of anionic and cationic surfactants as dispersant on impregnated MWCNTs/CS for the removal of CR dye.

The adsorption capacity of CNT-impregnated CSBs was found to be dependent on the nature of dispersant for CNTs. CNT-impregnated CSBs were prepared by four different strategies for dispersing CNTs: (a) in CS solution (CSBN1), (b) in sodium dodecyl sulfate (SDS) solution (CSBN2), (c) in CS solution contain in cetyl trimethyl ammonium bromide (CTAB) (CSBN3), and (d) in SDS solution for gelation with CTAB-containing CS solution (CSBN4). The adsorption capacities of composite for CR were in the order of: CSBN4 > CSBN1 > CSBN2 > CSBN3. The adsorption capacity of CSBN1 was slightly higher than CSBN2 for CR. This could be due to the dispersion in SDS molecules resulting in negatively charged CNTs as a result of adsorption of SDS molecules onto CNTs. The CR adsorption onto CSBN4 increased as the concentration of CNTs increased because of better dispersion of CNTs in SDS solution than in CTAB solution. CSBN3 showed poor adsorption capacity because CNTs were dispersed in CTAB and aggregates of CNTs blocked the adsorption sites on CS, CTAB, and SDS molecules in the beads [72].

However, covalent functionalization of CNTs with polymers has proved to be an effective way to improve their dispersion stability and make the resulting composites more stable and controllable CS, a natural polysaccharide with similar structural characteristics to cellulose, obtained by the deacetylation of chitin. It is a biocompatible, biodegradable, and non-toxic natural biopolymer and has excellent film-forming abilities. CS can selectively adsorb some metal ions, and has been successfully used in wastewater treatment. However, because CS is very sensitive to the pH of ionic solutions, its applications are limited. Several strategies have been devised to prepare CS derivatives that are insoluble in acid solutions and to preserve the adsorption capacity of CS. In order to increase the chemical stability of CS in acid solutions and improve its metal-ion-adsorbing properties, Schiff base-chitosan (S-CS) was produced by grafting aldehydes onto the CS backbone. Although the preparation of CS-modified MWCNTs via covalent interactions have been reported by a few groups, so far no work has been published on the application of MWCNTs covalently modified with CS derivative is to SPE column pre-concentration for ICP-MS determination of trace metals. A novel material was synthesized by covalently grafting S-CS onto the surfaces of MWCNTs, and used for pre concentration of V(V), Cr(VI), Cu(II), As(V), and Pb(II) in various samples, namely herring, spinach, river water, and tap water, using an SPE method. The results demonstrated that the proposed multi element enrichment method can be successfully used for analysis of V(V), Cr(VI), Cu(II),As(V), and Pb(II) in environmental water and biological samples. The method is fast and has good sensitivity and excellent precision. Compared with previously reported procedures, the present method has high enrichment factors and sensitivity. In short, the proposed method is suitable for pre concentration and separation of trace/ultra-trace metal ions in real samples.

1.1.6.2 CNT-ACF AS ADSORBENT

To resolve the aggregation and dispersion problem of CNTs, prepared the CNTs/ activated carbon fabric (CNTs/ACF) composite and its application was investigated for the removal of phenol and basic violet 10 (BV10). CNTs/ACF was prepared via directly growing nanoscaled CNTs on micro scaled carbon matrix. Poly acrylonitrile was used as a source of carbon. From the results, it was observed that dye adsorption equilibrium time for CNTs/ACF is shorter as compared to ACF and monolayer adsorption capacity does not display a linear increase with increasing the BET surface area. The decoration of CNTs tends to lower the porosity of the ACF. This finding indicates that the total microporosity of ACF cannot be fully accessed by the dye molecules. However, as-produced CNTs offer more attractive sites, including grooves between adjacent tubes on the perimeter of the bundles, accessible interstitial channels, and external nanotube walls. Therefore, the appearance of CNTs plays a positive role in (i) facilitating the pore accessibility to adsorbates and (ii) providing more adsorptive sites for the liquid-phase adsorption. This reflects that CNTs/ACF contains a large number of mesopore channels, thus preventing the pore blockage from the diffusion path of micropores for adsorbates to penetrate [72].

A novel technique was performed to grow high-density carbon nanotubes (CNTs) that attach to PAN-based ACFs. Such unique nano/microscaled carbon composites can serve as an excellent electrode material of EDLCs. The existence of CNTs is believed to play two important roles in enhancing the performance of EDLCs: (i) since its has good electric conductivity, the presence of CNTs would promote the contact electron transfer or lower contact resistance between the current collector and the carbon composite;(ii) CNTs not only provide additional exterior surface area for double-layer formation but also shift from micropore size distribution to mesopore size distribution that may reduce ionic transfer resistance and improve the high rated is charge capability. It is generally recognized that transition metals (Fe, Co, and Ni) can serve as catalytic sites in inducing carbon deposition, thus forming CNTs. This reveals that the uniform dispersion of the "seeds" on ACF surface, followed by a catalytic chemical vapor deposition (CCVD) treatment, would offer a possibility to fabricate the unique nano/microscaled carbon composites. The mesoporous channels, which came from CNT branches, would provide available porosity that is accessible for ionic transport and energy storage. This work intends to investigate the applicability of using the carbon composite as an electrode material for EDLCs. Two configurations (with and without CNTs) have been compared with respect to their double-layer capacitances and high-rate capability, analyzed by cyclic voltammetry (CV) and charge–discharge cycling. This work has demonstrated that the specific capacitance and high-rate capability of PAN-based ACF in H_2SO_4 can be enhanced with the decoration of CNTs. A CVD technique enabled to catalytically grow CNTs onto ACF, thus forming CNT–ACF composite. N_2 physisorption indicated that the mesopore fraction of ACF is found to increase after the growth of nanotubes. The specific double-layer capacitance and high-rate capability were significantly enhanced because of the presence of nanotubes. The distributed capacitance effect and inner resistance were significantly improved for the CNT–ACF capacitor.

Owing to the decoration of CNTs, the specific capacitance was found to have an increase of up to 42%.

1.1.6.3 CNT-METAL OXIDES AND MAGNETIC COMPOSITES AS ADSORBENT

Incorporation of magnetic property in CNTs is another good technique to separate CNTs from solution. The magnetic adsorbent can be well dispersed in the water and easily separated magnetically. Magnetic-modified MWCNTs were used for removal of cationic dye crystal violet (CV), thionine (Th), janus green B (JG), and methylene blue (MB). To find the optimum adsorption, effect of various parameters including initial pH, dosages of adsorbent and contact time have been investigated. The optimum adsorption was found to be at pH 7.0 for all dyes. In recent work they prepared guar gum grafted Fe_3O_4/MWCNTs (GG/MWCNT/ Fe_3O_4) ternary composite for the removal of natural red (NR) and methylene blue (MB). The removal efficiency of cationic dyes using GG/MWCNT/Fe_3O_4 is higher as compared with other adsorbents such as MWCNTs and MWCNT/Fe_3O_4. The higher adsorption capacity of GG/MWCNT/Fe_3O_4 could be related to the hydrophilic property of GG, which improved the dispersion of GG–MWCNT–Fe_3O_4 in the solution, which facilitated the diffusion of dye molecules to the surface of CNTs. The magnetic GG/MWCNT/Fe_3O_4possesses the high adsorption properties and magnetic separation and can therefore be used as magnetic adsorbents to remove the contaminants from aqueous solutions. Starch functionalized MWCTs/ iron oxide composite was prepared to improve the hydrophilicity and biocompatibility of MWCNTs. Synthesized magnetic MWCNT–starch–iron oxide was used as an adsorbent for removing anionic methyl orange (MO) and cationic methylene blue (MB) from aqueous solutions. MWCNT–starch–iron oxide exhibits super paramagnetic properties with a saturation magnetization and better adsorption for MO and MB dyes than MWCNT–iron oxide. The specific surface areas of MWCNT/iron oxide and MWCNT–starch–iron oxide were 124.86 and 132.59 m^2/g. However, surface area of the composite was small but adsorption capacity was higher compared to the parent one, the adsorption capacities of MB and MO onto MWCNTs–starch–iron oxide were 93.7 and 135.6 mg/g, respectively, while for MWCNTs–iron oxide were 52.1 and 74.9 mg/g. This is again confirming that ternary composite has the higher removal capacity than the binary composite and as grown CNTs. A novel magnetic composite bio adsorbent composed of chitosan wrapping magnetic nanosized γ-Fe_2O_3 and multi-walled carbon nanotubes (m-CS/γ-Fe_2O_3/MWCNTs) was prepared for the removal of methyl orange. The adsorption capacity of MO onto m-CS/γ-Fe_2O_3/MWCNTs was 2.2 times higher than m-CS/γ-Fe_2O_3. The adsorption capacity of MO onto m-CS/γ-Fe_2O_3/MW-CNTs (66.90 mg/g) was also higher than MWCNTs (52.86 mg/g). Kinetics data and adsorption isotherm data were better fitted by pseudo second-order kinetic model and by Langmuir isotherm, respectively. On the comparison of nature of adsorption of MO onto MWCNTs and m-CS/γ-Fe$_2$O$_3$/MWCNTs, observed that adsorption of MO onto MWCNTs was endothermic. CS is responsible for the exothermic adsorption process because with the increase in the temperature, polymeric network of CS changed/de-shaped, which reduced the porosity of the bio sorbent and hindered the diffusion of dye molecules at high temperature [72, 73].

A fast separation process is obtained by magnetic separation technology. Therefore, the adsorption technique with magnetic separation has aroused wide concern. However, the highest maximum adsorption capacity is not high enough. Polymer shows excellent sewage treatment capacity in the environmental protection. Recently reported that the decolorization efficiency could get close to 100% in the direct light resistant black G solution by P(AM-DMC) (copolymer of acrylamide and 2-[(methacryloyloxy) ethyl] trimethyl ammonium chloride). Therefore, the integration of CNTs, magnetic materials and polymer could overcome those defects of low adsorption capacity, long adsorption time, separation inconvenience and secondary pollution However, there were few reports about such nanocomposite. Herein, the first attempt to prepare the MPMWCNT nanocomposite with the aid of ionic liquid-based polyether and Ferroferric oxide was made. The structure of the nanocomposite was characterized and the physical properties were investigated in detail. Combined with the characteristics of the three materials, the magnetic nanocomposite shows excellent properties of short contact time, large adsorption capacity, rapid separation process and no secondary pollution in the adsorption process. This chapter could provide new adsorption insights in wastewater treatment. Then, the pH value of the final suspensions was adjusted to 10–11 and the redox reaction continued for 30 min with stirring. The MPMWCNT nanocomposite was separated by a permanent magnet and dried under vacuum. The yield of the nanocomposite was about 66.5% [83].

Earlier studies have indicated that magnetic carbon nanocomposites may show great application potential in magnetic data storage, for magnetic toners in xerography, Ferro fluids, magnetic resonance imaging, and reversible lithium storage. However, there is still lack of a systematic review on the synthesis and application of carbon-based/magnetic nanoparticle hybrid composites. So here we present a short review on the progress made during the past two decades in synthesis and applications of magnetic/carbon nanocomposites, and synthesis of magnetic carbon nanocomposites. During the last decade, much effort has been devoted to efficient synthetic routes to shape-controlled, highly stable, and well defined magnetic carbon hybrid nanocomposites. Several popular methods including filling process, template-based synthesis, chemical vapor deposition, hydrothermal/solvothermal method, pyrolysis procedure, sol–gel process, detonation induced reaction, self-assembly method, etc., can be directed at the synthesis of high-quality magnetic carbon nanocomposites. Examples for the applications of such materials include environmental treatment, microwave absorption, electrochemical engineering, catalysis, information storage, biomedicine and biotechnology [83].

1.1.6.4 CNT-DOLOMITE, CNT-CELLULOSE AND CNT-GRAPHENE AS ADSORBENTS

Recently developed a novel adsorbent by inserting MWCNTs into the cavities of dolomite for scavenging of ethidium the foam line CNTs/dolomite adsorbent. Adsorptions reached equilibrium within 30 min for the cationic dyes, acridine orange, ethidium bromide, and methylene blue while it was about 60 min for the anionic dyes, eosin B and eosin Y. Foam like ternary composite PUF/diatomite/dispersed-MWCNTs, gave the highest capacities for adsorption of these dyes, followed by PUF/agglomerated-MWCNTs, and then PUF/dispersed-MWCNTs. Langmuir adsorption isotherm was best fitted to the equilibrium data. The removal of methylene blue onto natural tentacle type wale gum grafted CNTs/cellulose beads was investigated by some researchers. The maximum adsorption of MB was observed at pH 5 and 150 min. Adsorption isotherm study revealed the monolayer adsorption at higher concentration and multilayer adsorption at lower concentration. The equilibrium adsorption capacity onto the adsorbent was determined to be 302.1 mg/g at pH 6.0 from Sips model. Pseudo first order kinetics gives the best-fitted results compared to the pseudo second order. From the results, it is evident that carboxylic group on the adsorbent plays the important role for the removal of MB as ionized to COO− at higher pH and bind with MB through electrostatic force [1]. Self-assembled cylindrical graphene–MCNT (G–MCNT) hybrid, synthesized by the one pot hydrothermal process was used as adsorbent for the removal of methylene blue in batch process. G–MCNT hybrid showed good performance for the removal of MB from aqueous solution with a maximum adsorption capacity of 81.97 mg/g. The kinetics of adsorption followed the pseudo second-order kinetic model and equilibrium data were best fitted to Freundlich adsorption isotherm. The adsorption capacity of G–MCNTs is much higher than MCNTs. Therefore, G–CNTs hybrid could be used as an efficient adsorbent for environmental remediation [72–73].

1.1.6.5 CNT-CF (PAN) AS ADSORBENT

In order to combine carbon nanotubes with carbon fibers, most studies report the direct synthesis of carbon nanotubes on carbon fibers by CVD (chemical vapor deposition) with special attention paid to the control of CNT length, diameter and density as well as arrangement and anchorage on carbon fibers. The approach, which was mostly investigated, consists in impregnating carbon fibers with a liquid solution of catalyst precursors (nickel, iron or cobalt nitrates, iron chloride) followed by CNT growth from carbonaceous gaseous precursors such as ethylene and methane. Carbon fibers can also be covered by catalyst particles (iron, stainless steel) using sputtering or evaporation techniques or even electrochemical deposition before introducing a gaseous carbon source (methane, acetylene) for CNT growth. In order to improve the CNT growth efficiency on carbon fibers, different modifications have been made such as addition of H_2S in the reactive gas phase, or the development of chemical vapor infiltration (CVI) technique, enabling us to improve the density of CNT and/or their distribution on the fibers. Based on the diffusion of catalyst particles into carbon, as reported in several papers, some studies report the deposition of layers on C fibers playing the role of barrier against catalyst diffusion. Carbon fiber coating with SiO_2-based thick layers through coprecipitation or solgel methods prior to CNT growth has been investigated, allowing the improvement of CNT density and distribution [4]. Growing aligned carbon nanotubes with controlled length and density is of particular importance for multi scale hybrid composites, since such nanotubes are expected to improve electrical and mechanical properties. Injection-CVD technique for both the deposition of the ceramic sublayer from organo metallic precursors and CNT growth from a hydrocarbon/metallocene precursor mixture. This process is efficient for the growth of aligned nanotubes on carbon substrates, but also on metal substrates such as stainless steel, palladium, or any metal substrates compatible with the CNT synthesis temperature.

The injection-CVD setup is quite similar to the one described by some researchers and has been modified in order to achieve CNT growth on carbon fibers and metal substrates. In particular, the evaporation chamber referred to later as 'the evaporator' was fitted with two identical injectors, derived from standard car engine injectors. A 1 M tetra ethyl ortho silicate (TEOS) solution in toluene was used as precursor for the deposition of silica sub layers. A ferrocene (2.5 wt. %) solution in toluene was used for the growth of CNTs. The tank feeding the first injector is filled with the sublayer precursor solution while the tank feeding the second injector is filled with the CNT precursor solution. For the deposition of the silica sublayers, the evaporator temperature is set to 220°C and the furnace temperature is set to 850°C. The Ar carrier gas flow rate is adjusted to 11 g min^{-1} and the pressure is regulated at 100 mbar. The sublayer precursor solution injection rate is 0.96 g min^{-1} TEOS solution was injected into the oven over 23–230 s. Once the sublayer deposition is realized, the evaporator temperature is lowered to 200°C and the Ar flow rate is adjusted to 31 min^{-1}. The injection of the ferrocene/toluene solution with a 0.75 g min^{-1} injection rate over 1–15 min takes place in the same furnace without manipulation of the pretreated fibers. In this experiments, CNT growth was performed over 1–10 min on carbon fibers covered with a SiO_2-based layer deposited over the longer duration and was compared to CNT growth on a flat quartz substrate, whereas in CNT growth experiment was performed over 15 min on carbon fibers covered with a SiO_2-based layer deposited over variable durations and was compared to CNT growth on a flat quartz substrate. In addition, some experiments were performed on quartz fibers in order to check the influence of the morphology of the substrate on CNT arrangement.

Carbon fibers are widely used as reinforcement in composite materials because of their high specific strength and modulus. Such composites have become a dominant material in the aerospace, automotive and sporting goods industries.

Current trends toward the development of carbon fibers have been driven in two directions; ultrahigh tensile strength fiber with a fairly high strain to failure (2%), and ultrahigh modulus fiber with high thermal conductivity. Today, a number of ultrahigh strength polyacrylonitrile (PAN)-based (more than 6 GPa), and ultrahigh modulus pitch-based (more than 900 GPa) carbon fibers have been commercially available. Carbon fibers with exceptionally high thermal conductivity are critical for many thermal control applications in the aerospace and electronics industries. The thermal conductivity of carbon fibers was found to increase asymptotically as the degree of preferred orientation of the crystalline parts in the fiber increases. However, further improvement of thermal conductivity over the existing highly oriented pitch-based carbon fiber while retaining the desired mechanical property has proven to be very challenging. One of the most effective approaches to further increase the thermal conductivity is to graft carbon nanotubes (CNTs) on the carbon fibers. CNTs have an extremely high thermal conductivity in the axial direction, and the thermal conductivities of multi-walled CNTs had been reported to be as high as 3000 W/m K. The grafting of CNTs on carbon fibers using chemical vapor deposition and electro deposition has been reported in the literature. Some researchers reported that the grafting of CNTs improves the mechanical properties and Weibull modulus of ultrahigh strength PAN-based and ultrahigh modulus pitch-based carbon fibers. The effect of grafting CNTs on the thermal conductivity of T1000GB PAN-based and K13D pitch-based carbon fibers were investigated (Fig. 1.3). Recently reported a method for the self-assembled fabrication of a single suspended amorphous carbon nanowire on a carbon- MEMS platform by electro spinning and pyrolysis of PAN and polymers. Here, we explore this technique's potential to fabricate the CNT/PAN composite nanofibers anchored to electrodes and thus, investigate the graphitic and electrical properties of single suspended CNT/carbon composite nanofibers. The conductivity of electro spun carbonized CNT/PAN nano fiber is measured at four different concentrations of MWCNT in the PAN electro spinning solution. In order to understand the structural changes that are responsible for the increase in conductivity of these nanofibers, micro Raman spectroscopy, X-ray diffraction (XRD) and high-resolution transmission electron microscopy (HRTEM) are used. Results indicate that the crystallinity and electrical conductivity of these composite nanofibers increase with increase in concentration of CNTs.

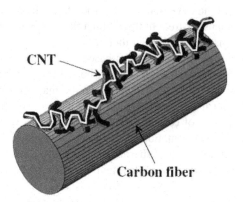

FIGURE 1.3   Schematic model of CNTs-grafted carbon fiber filament.

An effective strategy for positioning, integration and interrogation of a single nanofiber requires controlled electro spinning as detailed elsewhere. Researchers also determine the maximum concentration of CNTs in PAN that allows good electro spin ability of the precursor polymer to carbon nanowires. Overall fabrication methodology is a combination of three techniques: (1) photolithography to produce an MEMS structure, (2) self-assembled electro spinning of functionalized CNTs in PAN solution to form nanowires anchored on the MEMS platform, and (3) controlled pyrolysis to obtain carbon composite wires integrated with the underlying carbon MEMS structure. Four different concentrations of CNTs in PAN for electro spinning were prepared by mixing 0.05, 0.1, 0.2 and 0.5 wt.% CNTs in 8 wt.% PAN in DMF. The detailed method for solution preparation is described elsewhere. These solutions were electrospun using Dispovan syringe (volume: 2.5 mL and diameter: 0.55 mm) at 13–15 kV on the MEMS structures fabricated earlier. The distance between the tip of the jet and the MEMS collector was 10 cm, and the flow rate of the solution was maintained at 1 L/min Electro spinning was performed for a short period of 5–10 s to limit the number of wires and to obtain single wires suspended between posts. The resulting structures consisting of suspended composite nanowires on polymers were stabilized in air at 250°C for 1 h prior to pyrolyzing the whole structure which is done under N_2 flow (flow rate: 0.2 L/min) at 900°C for 1 h, with a ramp time of 5°C/min, to yield a monolithic carbon structure having good interfacial contacts and with electrically conducting posts of much greater cross-sectional area than the nanofibers themselves.

Carbon fibers used in this study are (i) a low thermal conductivity and ultrahigh tensile strength PAN-based (T1000GB) carbon fiber and (ii) a high thermal conductivity and ultrahigh modulus pitch-based (K13D) carbon fiber. Note that as received, both fibers had been subjected to commercial surface treatments and sizing (epoxy compatible sizing). To grow CNTs on the carbon fibers, an $Fe(C_5H_5)_2$ (ferrocene) catalyst was applied to the samples fiber bundles using thermal chemical vapor deposition (CVD) in vacuum. Prior to the application of the catalyst, the carbon fiber bundles were heat treated at 750°C for an hour in vacuum to remove the sizing. The growth temperature and time for CNTs deposition were selected as 750°C(T1000GB) and 700°C (K13D) for 900 s.

1.1.6.6 CNT-AC COMPOSITE

Among various advanced functional materials, electronically conducting polymers (such as polypyrrole (PPy) and polyaniline (PANI)) and metal oxides (such as RuO_2, MnO_2, NiO, and Co_3O_4) are widely used in super-capacitors. However, their applications are severely limited by their poor solubility and mechanical brittleness. Anchoring the conductive polymers or metal oxides to cellulose fibers or other textile fibers has inspired the design of their paper or textile based composites which show excellent cycling stability, mechanical flexibility and robustness. However, the textile fibers are usually insulators. Advanced carbon materials, such as carbon nanotubes (CNTs), graphene, ordered mesoporous carbon, carbon aerogels, hierarchical porous carbon, carbide-derived carbon, and their composites/hybrids, have been widely explored for use as super-capacitor electrodes. The carbon materials can also be combined with conductive polymers or metal oxides to obtain flexible CNT/PANI, carbon nanofiber (CNF)/PANI, graphene/PANI or graphene oxide/PANI, or CNT/CuO composite electrodes with improved electrochemical performance. Very recently, bi-scrolling nanotube sheets and functional guests into yarns, which contained up to 95 wt.% of otherwise unspinnable particulate or nanofiber powders, has been used to fabricate yarns for use in super-capacitors and lithium on battery materials. However, the large-scale production of inexpensive, flexible electrode remains a great challenge. Super-capacitors with AC/CNT nanocomposite electrodes have been shown to exhibit enhanced electrochemical performance compared with CNT-free carbon materials, although the original CNTs were strongly entangled with each other, and acid purification was always required. As a result, those CNTs well dispersed in the electrode were too short to form a self-supporting network. The as-obtained AC/CNT nanocomposites were still in powder form, and a binder was still needed. Recently it has been shown that vertically aligned CNTs, in which the CNTs with large aspect ratio are well oriented, can be well dispersed into individually long CNTs by a two-step shearing strategy. They also obtained CNT pulp, in which long CNTs have good dispersion in the liquid phase, can be used as a feedstock for CNT transparent conductive films and Bucky paper. As a result of great efforts to mass-produce aligned CNTs, they can be easily produced by radial growth on spheres or intercalated growth in lamellar catalysts. In this contribution, industrially produced aligned CNTs, together with AC powder, were used as raw materials to fabricate flexible electrodes. It is expected that the CNTs will bind AC particles together to give a paper like composite. 90 wt.%–99 wt.% of AC was first incorporated into the CNT pulp, and the deposited on a filter to make composites [72].

The specific surface area, pore volume and pore size distribution are also the important factors affecting the electrochemical performance of super-capacitors. The N_2 adsorption isotherms and pore size distributions of AC powder and CNTs are studied. The quantitative data extracted from the N_2 adsorption isotherms are given. AC presents a type- I isotherm, which is typical for a microporous material. The BET surface area and average pore diameter of AC powder were 1374 m^2/g and 2.35 nm, respectively. After the addition of AC or CNTs, the N_2 adsorption was almost unmodified in the small relative pressure region corresponding to micropores, whereas a noticeable enhancement at $P/P_0 > 0.9$, in the range of mesoporosity, was observed. The pore size distribution of CNTs showed two peaks at about 2.5 and 20 nm. The former may arise from the inner space of CNTs and the latter probably arise from the pores between CNTs. The pore size distribution of AC-CNT-5% also showed two small peaks at about 10 and 25 nm. The former may be generated by the pores in bundles of CNTs formed during the liquid phase process. The latter results from the pores generated by the stacking of CNTs, and its size was slightly increased due to the presence of AC particles. The BET surface area and average pore diameter of AC/AB were 1011 m^2/g and 2.64 nm. The total pore volume decreased due to the addition of 15 wt.% AB and 10 wt.% PTFE. However, the ratio between the mesoporous and microporous volumes (Vmeso /Vmicro) was almost the same. The BET surface area and average pore diameter of AC-CNT-5% were 1223 m^2/g and 2.99 nm. The micropore volume of the AC was slightly diminished by the presence of CNTs, whereas the mesoporous volume increased from 0.32 to 0.38 cm^3/g. The Vmeso/Vmicro ratio showed a significant increase.

1.1.6.7 CNT-FIBER COMPOSITES AS ADSORBENT

1.1.6.7.1   CNT-BASED FIBERS

For many applications, fibrous materials are more suitable than bulk materials. In addition, fiber production techniques tend to be suited for the alignment of nanotubes within the polymer matrix. The researchers observed that the alignment of nanotubes within the composite fiber was improved dramatically by increasing the draw ratio. Many subsequent studies also showed that the mechanical/electrical properties of these composite materials are greatly enhanced by fiber spinning. A part from the traditional melt spinning methods, composite fibers can also be produced by solution-based processing. The coagulation-spinning method was designed so that the CNTs were attached to each other while they were oriented in a preferential direction by a given flow. Nanotube aggregation was obtained by injecting the CNT dispersion into a rotating aqueous bath of PVA, such that nanotube and PVA dispersions flowed in the same direction at the point of injection. Due to the tendency of the polymer chains to replace surfactant molecules on the graphitic surface, the nanotubes dispersion was destabilized and collapsed to form a fiber. These wet fibers could then be retrieved from the bath, rinsed and dried. Significant rinsing was used to remove both surfactant and PVA. Shear forces during the flow lead to nanotube alignment. These fibers displayed tensile moduli and strength of 9–15 GPa and ~150 MPa, respectively. For stretched CNT/DNA/PVA fibers, the tensile moduli and strength were ~19 GPa and ~125 MPa, respectively. In the meantime, the coagulation-spinning method was further optimized by others. They injected the SWCNT dispersion into the center of a co-flowing PVA/water stream in a closed pipe. The wet fiber was then allowed to flow through the pipe before being wound on a rotating mandrel. Flow in more controllable and more uniform conditions in the pipe resulted in more stable fibers. Crucially, wet fibers were not rinsed to remove most of PVA (final SWCNT weight fraction ~60%). This resulted in large increases in Young's modulus and strength to 80 and 1.8 GPa, respectively. Furthermore, study works have shown that single- and multi-walled CNTCNT based fibers could be drawn at temperatures above the PVA glass transition temperature (~180°C), resulting in improved nanotube alignment and polymer crystallinity. These so-called hot-stretched fibers exhibited values of elastic moduli between 35 and 45 GPa and tensile strengths between 1.4 and 1.8 GPa, respectively. In an alternative approach, CNT/polymer solutions have been spun into fibers using a dry-jet wet spinning technique. This was achieved by extruding a hot CNT/polymer solution through a cylindrical die. An approximately 1–10 cm air gap was retained between the die orifice and the distilled water coagulation bath, which was maintained at room temperature. Significant mechanical property increases were recorded for the composite fibers compared with the control samples with no CNT reinforcement. Another method used recently to form composite-based fibers from solution is electro spinning. This technique involves electro statically driving a jet of polymer solution out of a nozzle onto a metallic counter electrode. In 2003, two groups independently described electro spinning as a method to fabricate CNT–polymer composite fibers. Composite dispersions of CNTs in either PAN or PEO in DMF and ethanol/water, respectively, were initially produced. Electro spinning was carried out using air pressure of 0.1–0.3 kg/cm² to force the solution out of a syringe 0.5 mm in diameter at a voltage difference of 15–25 kV with respect to the collector. Charging the solvent caused rapid evaporation resulting in the coalescence of the composite into a fiber, which could be collected from the steel plate. Fibers with diameters between 10 nm and 1 m could be produced in this fashion.

1.1.6.7.2   TEXTILE ASSEMBLIES OF CNTS

Some researchers have demonstrated the feasibility in processing of nanotube yarns with high twist spun from nanotube forests, plied nanotube yarn processed from a number of single nanotube yarns with counter direction twist and 3-D nanotube braids fabricated from 365-ply yarns. In addition, the plied nanotube yarns and 3-D braids were used as through-the-thickness (Z) yarns in 3-D weaving process. The challenges in making nanotube fibers/yarns with desirable properties, according to recent work, are in achieving the maximum possible alignment of the nanotubes or their bundles within the yarn, increasing the nanotube packing density within the yarn and enhancing the internal bonding among the nanotubes. Following this approach then produced MWCNT single-ply yarns and 5-ply yarns, which were made by over twisting five single yarns and subsequently allowing them to relax until reaching a torque-balanced state. First, the CVD synthesized MWCNTs are 300 mL long and 10 nm in diameter and they formed about 20 nm diameter bundles in a nanotube forest. Simultaneous draw and twist of the bundles produced 10 mL diameter single yarn. Besides single-ply and 5-ply yarns, Many Researchers have also demonstrated the fabrication of other multiply yarns which have been used in 3-D braiding process as well as Z-yarns in 3-D weaving processes. It is currently possible to produce tens of meters of continuous MWCNT yarns. It is reported that no visible damage to the nanotube yarns is imparted by the braiding process and the 3-D braids are very fine, extremely flexible, hold sufficient load, and are well suited for the use in any other textile formation process, or directly as reinforcement for composites. The reported elastic and strength properties of carbon nanotube composites so far are rather low in comparison with conventional continuous carbon fiber composites. It is believed that the properties can be substantially improved if the processing methods and structures are optimized. Others also studied the electrical conductivity of CNT yarns, 3-D braids and their composites. It is noted that 3-D textile composites, including 3-D woven and 3-D braided materials, combine high in-plane mechanical properties with substantially improved transverse strength, damage tolerance and impact resistance. However, even relatively small volume content of the out-of-plane fibers results in considerable increase of interstitial resin pockets, which contributes to lower in-plane properties. Utilizing fine CNT yarns could dramatically reduce the through the-thickness yarn size while still sufficiently improving the composite transverse properties. Furthermore, they found that the electrical conductivity of 3-D hybrid composites are many times greater than that of commonly produced nano composites made from low volume fraction dispersion of relatively short CNTs in epoxy.

1.1.6.7.3   CARBON NANOTUBE FIBERS

The superb mechanical and physical properties of individual CNTs provide the impetus for researchers in developing high-performance continuous fibers basedupon carbon nanotubes. Unlike in the case of carbon fibers, the processing of CNT fibers does not require the cross-linking step of the precursor structures. As summarized results, the leading approaches for the production of CNT fibers are (1) spinning from a lyotropic liquid crystalline suspension of nanotubes, in a wet-spinning process similar to that used for polymeric fibers such as aramids, (2) spinning from MWCNTs previously grown on a substrate as "semi aligned" carpets and (3) spinning directly from an aerogel of SWCNTs and MWCNTs as they are formed in a chemical vapor deposition reactor. These methods as well as the twisting of SWCNT film are reviewed. Recently, some researchers reported the properties of epoxy/CNT fiber composites, which are similar to composites reinforced with commercial carbon fibers. The composites were formed by the diffusion of uncured epoxy into an array of aligned fibers of CNTs. The tensile and compressive properties were measured. The results demonstrated significant potential of CNT fiber reinforced composites. The work also highlights the issue in defining the cross-section of CNT fibers and other CNT macro-assemblies for mechanical property evaluation. Mora et al. noted that the volumetric density of CNT fibers in epoxy composites was found to be much higher than the density of the as-spun fiber obtained from gravimetric and diameter measurements. This difference is related to the volume of free space inside a bundle of fibers. This space is infiltrated by the epoxy and ultimately increases the CNT/polymer interface area. The catalyst particles must be eliminated from the fiber; drawing conditions must be optimized to eliminate entanglements between CNTs; and the fiber needs to be pulled at the rate at which nanotubes are growing so the growth of an individual nanotube is not terminated.

1.1.6.7.4   GEL-SPINNING OF CNT/POLYMER FIBERS

CNTs can act as a nucleation agent for polymer crystallization and as a template for polymer orientation. SWCNTs with their small diameter and long length can act as ideal nucleating agents. Study works suggested that the next-generation carbon fibers will likely be processed not from polyacrylo nitrile alone but from its composites with CNTs. Furthermore, continuous carbon fibers with perfect structure, low density, and tensile strength close to the theoretical value may be feasible if processing conditions can be developed such that CNT ends, catalyst particles, voids, and entanglements are eliminated. Such a CNT fiber could have ten times the specific strength of the strongest commercial fiber available today. In the current manufacture process of carbon fibers, polymer solution is extruded directly into a coagulation bath. However, higher strength and modulus PAN and PAN/SWCNT based fibers can be made through gel spinning. In gel spinning, the fiber coming out of the spinneret typically goes into a cold medium where it forms a gel. These gel fibers can be drawn to very high draw ratios. Gel-spun fibers in the gel bath are mostly unoriented and they are drawn anywhere from 10 to 50 times after gelation. Structure of these fibers is formed during this drawing process. The gel-spinning process, invented around 1980 has been commercially practiced for polyethylene to process high-performance fibers such as Spectra™ and Dyneema™. Although small diameter PAN fibers (10 nm to 1 mL diameter) can be processed by electro spinning, the molecular orientation and hence the resulting tensile modulus achieved is rather low, and processing continuous fiber by electro spinning has been challenging. Others have used island-in-a-sea bi-component geometry along with gel-spinning to process PAN and PAN/CNT fibers to make carbon fibers with effective diameters less than 1 mL. Small-diameter fibers possess high strength and gel-spinning results in high draw ratio and consequently high orientation and modulus.

1.1.6.7.5   ELECTROSPINNING OF CNT/POLYMER FIBERS

Electrospinning is an electrostatic induced self-assembly process, which has been developed for decades, and a variety of polymeric materials have been electro spun into ultra-fine filaments [22]. Electro spinning of CNT/polymer fibrils is motivated by the idea to align the CNTs in a polymer matrix and produce CNT/ polymer nanocomposites in a continuous manner. The alignment of CNTs enhances the axial mechanical and physical properties of the filaments. Recently researchers have adopted the coelectro spinning technique for processing CNT/ PAN (polyacrylonitrile) and GNP (graphite nanoplatelet)/PAN fibrils. The fluid is contained in a lass syringe, which has a capillary tip (spinneret). When the voltage reaches a critical value, the electric field overcomes the surface tension of the suspended polymer and a jet of ultra-fine fibers is produced. As the solvent evaporates, a mesh of nano to micro size fibers is accumulated on the collection screen. The fiber diameter and mesh thickness can be controlled through the variation of the electric field strength, polymer solution concentration and the duration of electro spinning. In the processing of CNT/PAN nanocomposite fibrils, polyacrylonitrile with purified high-pressure CO disproportionation (HiPCO) SWCNTs dispersed in dimethyl formamide, which is an efficient solvent for SWCNTs, are coelectron spun into fibrils and yarns. CNT-modified surfaces of advanced fibers prepared first time as modified the surface of pitch-based carbon fiber by growing carbon nanotubes directly on carbon fibers using chemical vapor deposition. When embedded in a polymer matrix, the change in length scale of carbon nanotubes relative to carbon fibers results in a multiscale composite, where individual carbon fibers are surrounded by a sheath of nanocomposite reinforcement. Single-fiber composites have been fabricated to examine the influence of local nanotube reinforcement on load transfer at the fiber/matrix interface. Results of the single-fiber composite tests indicate that the nanocomposite reinforcement improves interfacial load transfer. Selective reinforcement by nanotubes at the fiber/matrix interface likely results in local stiffening of the polymer matrix near the fiber/matrix interface, thus, improving load transfer. The interfacial shear strength of CNT coated carbon fibers in epoxy was studied using the single-fiber composite fragmentation test. Randomly oriented MWCNTs and aligned MWCNTs coated fibers demonstrated a 71% and 11% increase in interfacial shear strength over unsized fibers. Researchers attributed this increase to the increase in both the adhesion of the matrix to the fiber and the interphase shear yield strength due to the presence of the nanotubes. Another method to exploit the axial properties of CNTs is to assemble them into a macroscopic fiber, with the tubes aligned parallel with the fiber axis; a strategy similar to that proposed eight decades ago for the development of high-performance polymer fibers. Carbon nanotube fibers can be produced by drawing from an array of vertically aligned CNTs, by wet-spinning from a liquid crystalline suspension of CNTs or they can be spun directly from the reactor by drawing them out of the hot-zone during CNT growth by chemical vapor deposition (CVD). Considerable attention has been devoted to optimizing the structure of CNT fibers at the different stages of their production: controlling the synthesis of specific nanotubes, the assembly of CNTs into a fully dense fiber and using post-spin treatments to obtain specific properties. However, the integration of CNT fibers into composites and the properties of these composites have received comparatively less attention, in spite of these aspects being fundamental for many potential applications of this new high-performance fiber. A previous study on the mechanical properties of CNT fiber/epoxy composites showed effective reinforcement of the thermoset matrix both in tension and compression without the need for additional treatments on the CNT fiber. Large increases in stiffness, energy absorption, tensile strength and compressive yield stress were observed for composite with fiber mass fractions in the range 10–30%. CNT fibers have an unusual yarn-like structure with an accessible surface area several orders of magnitude higher than that of a traditional fiber. The free space between bundles in the CNT fiber is able to take up non- cross-linked resins by capillary action, which wick into the fiber and appear to fill the observable free space. As a consequence, the composite develops a hierarchical structure, with each CNT fiber, being an infiltrated composite itself. Measurements of CTE of CNT fiber composites show good stress transfer between matrix and fiber due to the good adhesion of the thermoset to the porous fibers aided by a significant level of structural 'keying.' The adding polymer into the CNT fibers does not disrupt the CNT bundle network; hence the electrical conductivity of the fiber is largely preserved and the electrical conductivity of the composite is increased. On the other hand, the thermal conductivity of the CNT fiber composite increases more rapidly than before with added fiber loading up to the maximum used here of 38%. Others suggest that the infiltration of the polymer into the fiber improves the thermal coupling between the nanotube bundles by filling the spaces between them, which is a characteristic of the as-spun condition. Addition of CNT fiber to the matrix produces an effective increase in composite thermal conductivity of 157 W/m K per unit fiber mass fraction. Results show easy integration of CNT fibers into axially aligned composites and rather effective exploitation of the electrical, thermal and mechanical axial properties of the CNT fibers thanks to the infiltration of polymer into them.

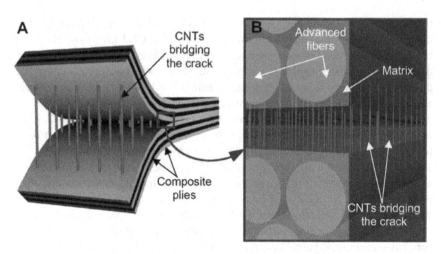

FIGURE 1.4 Illustration of the ideal hybrid interlaminar architecture: (A) CNTs placed in between two plies of a laminated composite and (B) close-up of the crack, showing CNTs bridging the crack between the two plies.

1.1.6.8 CNT-PANI COMPOSITES AS ADSORBENT

This study confirms that DP/MWCNTs is a potential adsorbent for the removal of Cr(VI) from aqueous solution. The functional groups present on the surface of adsorbent mainly, amine or imine groups of polyaniline and exposed surface and internal porous structure sites were mainly responsible for the adsorption of Cr(VI). The optimum Cr(VI) adsorption was found at pH 2. The equilibrium isotherm and kinetic results suggested that the adsorption was heterogeneous, multilayer and controlled through a boundary layer effect. The higher value of enthalpy change (DH) and incomplete desorption suggested that chemisorptions was main mechanism involved in the adsorption of Cr (VI).

Due to its outstanding stability in acidic solution, organic magnetic materials have attracted considerable attention and are viewed as a promising substitute for metal and metal oxide based magnetic materials. Among them, polyaniline (PANI) is one of the most attractive magnetic polymers because of its excellent chemical stability, easy synthesis and excellent magnetic property. Because the large amount of amine and imine functional groups of PANI are expected to have strong affinity with metal ions, PANI can enrich and remove heavy metal contaminants from aqueous solutions easily and effectively. Considering the promising magnetic property of PANI and excellent adsorption capacity of MWCNTs, PANI/MWCNTs magnetic composites might be an attractive material in the removal of heavy metal contaminants from large volumes of aqueous solutions. Comparing to conventional chemical methods, plasma technique has many advantages in surface modification of various materials. Some researchers also modified single walled carbon nanotubes with PANI by using plasma induced polymerization technique and studied its electrical properties. Herein, MWCNTs were modified with PANI by using plasma induced polymerization technique to synthesize PANI/MWCNTs magnetic composites. The prepared PANI/MWCNTs magnetic composites were characterized by ultraviolet–visible (UV–vis) spectrophotometer, X-ray photoelectron spectroscopy (XPS), thermo gravimetric analysis–differential thermal analysis TGA–DTA), field-emission scanning electron microscopy (FE-SEM), and vibrating sample magnetometer (VSM). The prepared PANI/MWCNTs magnetic composites were applied to adsorb and to enrich Pb(II) from aqueous solutions to evaluate its application in the removal of heavy metal contaminants from large volumes of aqueous solutions in environmental pollution cleanup. In the application of PANI/MWCNTs to remove Pb(II) ions in environmental pollution management, the pH value of aqueous solutions is also crucial because the adsorption of metal ions on solid particles is generally affected by the pH values. Moreover, the amine and imine functional groups of PANI and the oxygen containing functional groups of MWCNTs on PANI/MWCNT surfaces are protonated at low pH. The functional groups of PANI/MWCNTs are protonated easily at low pH, causing PANI/MWCNTs carry positive charges. Moreover, at low pH, the functional groups on PANI/MWCNT surfaces are competitively bound by the protons in aqueous solutions, which can restrict the enrichment of Pb(II) onto PANI/MWCNTs. The protonation of functional groups on PANI/MWCNT surfaces decreases with increasing pH, which results in the less positively charged PANI/MWCNTs. It is reasonable that the competitive binding for the functional groups on PANI/MWCNT surfaces between protons and Pb(II) will decrease. The removal of Pb(II) on MWCNTs and PANI/MWCNTs as a function of pH is dependent on pH values. The effect of ionic strength on the sorption of Pb (II) to PANI/MWCNTs is also important because of the presence of different cations and anions in the environment and the salt concentration in wastewater may be different for different sites. PANI/MWCNT magnetic composites were synthesized by using plasma induced polymerization technique. The analysis results of UV–vis spectra, XPS, TGA, and FE-SEM characterizations indicate that PANI has been modified onto MWCNTs. PANI/MWCNTs magnetic composites have very high adsorption capacities in the removal of Pb(II) ions from large volumes of aqueous solutions, and PANI/MWCNTs magnetic composites can be separated and recovered from solution by simple magnetic separation. The results of this research highlight the potential application of the magnetic composites of PANI/MWCNTs for heavy metal contaminants cleanup in the natural environment.

1.1.6.9 CNT-ALGINATE COMPOSITES AS ADSORBENT
Widespread usage of CNTs will cause increasing emissions to the water environment and result in human contact risk to CNTs.Another adverse factor currently restricting the application of CNTs to environmental protection/remediation is their high costs. These difficulties encountered in environmental applications

with alginate and CNTs can be overcome by formation of calcium alginate/multi-walled carbon nanotubes (CA/MWCNTs) composite fiber. As defined in our laboratory, CA/MWCNT composite fiber was prepared using $CaCl_2$ as cross-linking agent by wet spinning. The composite fiber not only makes full use of the heavy metals and dyes adsorption properties of alginate and MWCNTs, but also prevent MWCNTs from breaking off the composites to cause second micro pollution to water. Such technique is also a practical approach to overcome the high cost difficulty encountered in the use of CNTs for environmental remediation. Inrecent work, the adsorption properties for methylene blue (MB) and methyl orange (MO) organic dyes on CA/MWCNT composite fiber were characterized by using batch adsorption method. For preparation of SA/MWCNT dispersions, usually SA was dissolved in distilled water to produce a viscous solution with the concentration of 3 wt.% SA after mechanical agitation for 4 h at ambient temperature and continue stirring for30 min at 50°C. Then different amount of MWCNTs were added into SA solution under constant stirring for 30 min at room temperature and then ultrasonic 15 min to achieve homogeneous dispersion. Preparation of CA/MWCNT composite fibers Spinning dope was prepared for different concentrations of nanotube and SA. A narrow jet of spinning solution was injected through a 0.5 mm diameter needle into a coagulation bath containing a 5 wt.% aqueous solution of $CaCl_2$, and then collected on a spindle outside the bath, which was rotated. The coagulated fibers were then washed several times with deionized water and then dried in air under tension. The CA/MWCNT biocomposite fibers obtained had a linear-density of about 50 dtex. A novel material, the CA/MW-CNTs composite fiber as an effective bio adsorbent for ionic dyes removal has been prepared. The adsorbent toward ionic dyes has higher adsorption capacity due to the introduction of MWCNTs and the large specific area of fiber form of the composite. Batch adsorption experiments showed that the adsorption process followed second-order kinetic model. The results suggested that the initial pH value is one of the most important factors that affect the adsorption capacity of the dyes onto CA/MWCNTs. The desorption experiments showed the percentage of the desorption were found to be 79.7% and 80.2% for MB at pH2.0 and MO at pH 13.0, respectively, which were corresponding to the adsorption experiments. The CA/MWCNT composite fiber can be used as environment friendly bio adsorbent for the removal of ionic dyes from aqueous solution due to the efficient and fast adsorption process.

One effective method to resolve the second pollution caused by CNTs is to search for suitable supporters to immobilize CNTs for preparing macroscopic CNTs composites in order to make full use of the current microsized CNTs. Alginate, the salt of alginic acid, has hydrophilicity, biocompatiblity, non-toxicity, exceptional formability and is a linear chain structure of guluronic acid (G) residues arranged in a block wise fashion. It has a high affinity and binding capacity for metal ions and has already been widely used as heavy metal adsorbent in environmental protection. Some researchers reported that the adsorption iso-therms of UO_2^{2+} adsorbed by calcium alginate beads exhibited Langmuir behavior and its maximum monomolecular capacity reached 400 mg/g for UO_2^{2+} at 25°C. Others investigated Cu^{2+} removal capability of calcium alginate encapsulated magnetic sorbent and found that the maximum sorption capacity of the mate-rial can reach 60 mg/g at pH of 5.0 and temperature of 20°C and too prepared composite gels of calcium alginate containing iminodiacetic type resin. The bio-sorption capacity of the composite gels is higher than that of simple alginate gels and increases with increasing the amounts of resin enclosed in the composite. The adsorbents were characterized using SEM micrograph, BET surface analysis, FTIR spectra and Boehm titration method. The CA and CNTs/CA were depos-ited on a brass hold and sputtered with a thin coat of gold under vacuum, their morphology and surface CNTs. The results suggest that the higher surface area and pore volume of CNTs maybe act as microchannels in matrix and benefit for the improvement of surface area of the CNTs/CA composites. In summary, an ef-ficient adsorbent of CNTs/CA with exceptional Cu(II) adsorption capability was prepared. Equilibrium data were fitted to Langmuir and Freundlich isotherms. Based on Langmuir isotherm, the maximum monolayer adsorption capacity for CNTs/CA is 84.88 mg/g. The new type of adsorbent of CNTs/CA can resolve the micropollution problem caused by nano-sized CNTs through immobilizing them by CA and it will promote the practical applications of CNTs and their composites in environmental protection.

1.1.7 CARBON NANOFIBER (CNF): PROPERTIES AND APPLICATION

Carbon nanofibers (diameter range, 3–100 nm) have been known for a long time as a nuisance that often emerges during catalytic conversion of carbon-containing gases. The recent outburst of interest in these graphitic materials originates from their potential for unique applications as well as their chemical similarity to fuller-enes and carbon nanotubes. In this review, focused on the growth of nanofibers using metallic particles as a catalyst to precipitate the graphitic carbon. First, sum-marized some of the earlier literature that has contributed greatly to understand the nucleation and growth of carbon nanofibers and nanotubes. Thereafter, described in detail recent progress to control the fiber surface structure, texture, and growth into mechanically strong agglomerates. It is argued that carbon nanofibers are unique high surface area materials (200 mL/g) that can expose exclusively either basal graphite planes or edge planes. It is shown that the graphite surface structure and the lyophilicity play a crucial role during metal emplacement and catalytic use in liquid phase catalysis. An article by Iijima that showed that carbon nano-tubes are formed during arc-discharge synthesis of C_{60}, and other fullerenes also triggered an outburst of the interest in carbon nanofibers and nanotubes. These nanotubes may be even single walled, whereas low temperature, catalytically grown tubes are multi-walled. It has been realized that the fullerene type materials and the carbon nanofibers known from catalysis are relatives and this broadens the scope of knowledge and of applications. It has been realized, however, that arc-discharge and laser-ablation methods lead to mixtures of carbon materials and thus to a cumbersome purification to obtain nanofibers or nanotubes [88].

From an application point of view, Some of best application of carbon nano fibers include: Carbon nanofibers as catalyst support materials, Carbon nanofiber based electrochemical biosensors,CNF-based oxidase biosensors, CNF-based immune sensor and cell sensor and hydrogen storage. The overall economics are affected by the fiber yield, the feedstock used, the rate of growth, and the reactor technology [88–92].The growth of parallel fibers using iron as the catalyst has been studied in detail by high-resolution transmission electron microscopy (HR-TEM). It is noted that the graphite layers grow at an angle iron surface, thus leading to parallel fibers. The diameter of the fibers can be varied by variation of the metal particle size. If we want to vary the fiber diameter for a macroscopic sample, however, we need a narrow metal particle size distribution. In general, one can say that the fibers do not contain micropores and that the surface area can range from 10 to 200 m^2/g and the mesopore volume ranges between 0.50 and 2.0 mL/g. Note that these pore-volume data are obtained htiw fibers as grown, specific treatments in the liquid phase can be applied to largely reduce the pore volume and to obtain much denser and compact fiber structures. Compared to the large volume of literature on the mechanism of growth, the studies on the macroscopic, mechanical properties of bodies consisting of agglomerates of carbon nanofibers have been limited in number. Give a useful description of the tertiary structures that can be obtained; that is, "bird nests,""neponet," and "combedyarn." In general, porous bodies of carbon nanofibers are grown from porous supported metal catalyst bodies. Some others in the size range of micrometer to millimeter. As carbon precursors, PAN and pitches were frequently used, probably because both of them are also used in the production of commercial carbon fibers. In addition, poly(vinyl alcohol) (PVA), polyimides (PIs), polybenzimidazol (PBI) poly(vinylidene fluoride) (PVDF), phenolic resin and lignin were used. In order to convert electrospun polymer nanofibers to carbon nanofibers, carbonization process at around 1000°C has to be applied. In principle, any polymer with a carbon backbone can potentially be used as a precursor. For the carbon precursors, such as PAN and pitches, so-called stabilization process before carbonization is essential to keep fibrous morphology, of which the fundamental reaction is oxidation to change resultant carbons difficult to be graphitized at high temperatures as 2500°C [89, 90].

Carbon is an important support material in heterogeneous catalysis, in particular for liquid-phase catalysis. A metal support interaction between Ru and C was suggested as a possible explanation for these very interesting observations. More recent work focuses on the use of platelet type fibers exposing exclusively graphite edge sites. Using a phosphorus-based treatment, preferential blocking of so-called armchair faces occurs. Deposition of nickel onto the thus modified

CNF enabled one to conclude that the nickel particles active for hydrogenation of light alkenes reside on the zigzag faces. More characterization work is needed to substantiate these interesting claims. Others have carried out, by far, the most extensive work on CNF as carbon support material. A driving force for exploring CNF supports was related to the replacement of active carbon as support for liquid phase catalysis. For the CNF support, no shift of the PSD is apparent, whereas with an activated carbon (AC) support, severe attrition is apparent. They compare the PSD after ultrasonic treatment of the CNF and of the AC support. Clearly, AC displays a much broader PSD with, moreover, a significant number of fines. Nano composite electrodes made of carbon nanofibers and paraffin wax were characterized and investigated as novel substrates for metal deposition and stripping processes. Since CNFs have a much larger functionalized surface area compared to that of CNTs, the surface active groups to volume ratio of these materials is much larger than that of the glassy-like surface of CNTs. This property, combined with the fact that the number and type of functional groups on the outer surface of CNFs can be well controlled, is expected to allow for the selective immobilization and stabilization of bio molecules such as proteins, enzymes, and DNA. Additionally, the high conductivity of CNFs seems to be ideal for the electrochemical transduction. Therefore, these nano materials can be used as scaffolds for the construction of electrochemical biosensors [64–68].

Compared with conventional ELISA-based immunoassays, immune sensors are of great interest because of their potential utility as specific, simple, label-free and direct detection techniques and the reduction in size, cost and time of analysis. Due to its large functionalized surface area and high surface-active groups-to-volume ratio. Hydrogen storage is an essential prerequisite for the widespread deployment of fuel cells, particularly in transport. Huge hydrogen storage capacities, up to 67%, were reported. Unfortunately such astonishing values could not confirmed by other research teams worldwide. Volumetric and gravimetric hydrogen density for different storage methods, some reviews provide basics of hydrogen storage on carbon materials, the types of carbon materials with potential for hydrogen storage, the measured hydrogen storage capacities of these materials, andbased on calculations, an approximation of the theoretical achievable hydrogen storage capacity of carbon materials [91, 92].

1.1.8 ACTIVATED CARBON NANOFIBERS (ACNF): PROPERTIES AND APPLICATION

Electro spinning, a simple approach to make very fine fibers ranging from nano to micro scales, is attracting more attention due to the high porosity and high surface area to volume ratio of electros pun membranes. These properties contribute to potential applications of electros pun membranes in carbon and graphitic nanofiber manufacturing, tissue scaffolding, drug delivery systems, filtration and reinforced nanocomposites. The researchers also tried poly (amic acid) as a precursor to make activated carbon nano fibers. These studies employed physical activation to produce pores in precursor fibers. Compared to physical activation, the chemical activation process has important advantages, including low heat treatment temperature, short period of processing time, large surface area and high carbon yield; however, there has been no work reported for chemicallyactivated carbon nanofibers from electros pun PAN. The effects of electros pinning variables, such as applied voltage, pump flow rate and distance between the needle tip and collector, on the resulting nanofiber diameters were studied. Mechanical properties, such as tensile, tear and burst strength, of electros pun PAN non-woven membranes were measured and the quantitative relationships between membrane thickness and these properties were established [93–95]. Other physical properties, such as air permeability, inter fiber pore size and porosity, were also studied. Activated carbon nanofibers were produced from electrospun PAN by chemical activation with potassium hydroxide (KOH) as the activating agent. They were characterized by morphology, Fourier transforms infrared spectroscopy (FTIR), Brunauer-Emmett-Teller (BET) surface area, total pore volume and pore size distribution. There are two processes for manufacturing the carbon nanofiber (CNF), namely, the vapor-grown approach and the polymer spinning approach. The activated carbon nanofiber (ACNF) is the physically or chemically activated CNF, which have been, in many investigations, practically applied in electric double layer capacitors, organic vapor recovery, catalyst support, hydrogen storage, and so on. In practice, the physical activation method involves carbonizing the carbon precursors at high temperatures and then activating CNF in an oxidizing atmosphere such as carbon dioxide or steam [96, 97].

The chemical activation method involves chemically activating agents such as alkali, alkaline earth metals, some bases such as potassium hydroxide (KOH) and sodium hydroxide, zinc chloride, and phosphoric acid (H_3PO_4). In essence, most chemical activation on CNF used KOH to get highly porous structure and higher specific surface area. Unfortunately, large amount of solvent were needed to prepare the polymer solution for electros pinning and polymer blend, causing serious environmental problem thereafter. A series of porous amorphous ACNF were studied. Utilizing the core/shell microspheres, which were made of various polymer, blends with solvent. In their approach, the phenol formaldehyde-derived CNF were chemically activated by the alkaline hydroxides, and the thus-prepared ACNF were applied as super-capacitor electrodes and hydrogen storage materials [98–100].

In a continuous effort, some researchers proceed to investigate and compare the various chemical activation treatments on the CNF thus prepared, with particular emphasis on the qualitative description and quantitative estimation on the surface topology by AFM, and their relation to the microstructure of ACNF.PAN fiber following the spinning process by several ways such as modification through coating, impregnation with chemicals (catalytic modification) and drawing/

stretching with plasticizer. The post spinning modifications indirectly affect and ease the stabilization in several ways such as reducing the activation energy of cyclization, decreasing the stabilization exotherm, increasing the speed of cyclization reaction, and also improving the orientation of molecular chains in the fibers. One of the well-known posts spinning treatment for PAN fiber precursor is modification through coatings. The PAN fibers are coated with oxidation resistant resins such as lubricant (finishing oil), antistatic agents, andemulsifiers, which are basically used as, spin finish on the precursor fiber. Coating with certain resins also acts in the same manner as the comonomer in reducing the cyclization exotherm thus improving the mechanical properties of the resulting carbon fibers. Due to their excellent lubricating properties, silicone based compounds are mostly used as the coating material for PAN precursor fibers. Tensile load and tear strength of electrospun PAN membranes increased with thickness, accompanied with a decrease in air permeability; however, burst strength was not significantly influenced by the thickness [101–103].

Electrospun PAN nanofiber membranes were stabilized in air and then activated at 800°C with KOH as the activating agent to make activated carbon nanofibers. Stabilized PAN membranes showed different breaking behaviors from those before stabilization. The activation process generated micropores which contributed to a large surface area of 936.2 m^2/g and a micropore volume of 0.59 cc/g. Pore size distributions of electrospun PAN and activated carbon nanofibers were analyzed based on the Dubinin-Astakhov equation and the generalized Halsey equation. The results showed that activated carbon nanofibers had many more micropores than electrospun PAN, increasing their potential applications in adsorption. Based on a novel solvent-free coextrusion and melt-spinning of polypropylene/(phenol formal dehydepolyethylene)-based core/sheath polymer blends, a series of activated carbon nanofibers (ACNFs) have been prepared and their morphological and microstructure characteristics analyzed by scanning electron microscopy, atomic force microscopy (AFM), Raman spectroscopy, and X-ray diffractometry with particular emphasis on the qualitative and quantitative AFM analysis. Post spinning treatment of the current commercial PAN fiber based on the author's knowledge, reports on the post spinning modification process of the current commercial fiber are still lacking in the carbon fiber manufacturers' product data sheet, which allows us to assume that the current commercial carbon fibers still do not take full advantage of any of these treatments yet. During stabilization and carbonization of polymer nanofibers, they showed significant weight loss and shrinkage, resulting in the decrease of fiber diameter [104–107].

From an application point of view, some of best application of carbon nanofibers include: ACNF as anodes in lithium-ion battery, Organic removal from waste water using, ACNF as cathode catalyst or as anodes for microbial fuel cells (MFC), Electrochemical properties of ACNF as an electrode for super-capacitors, Adsorption of some toxic industrial solutions and air pollutants on ACNF

[108–120]. Activated carbon nano fibers (ACNFs) with large surface areas and small pores were prepared by electros pinning and subsequent thermal and chemical treatments. These activated CNFs were examined as anodes for lithium-ion batteries (UBs) without adding any non-active material. Their electrochemical behavior show improved lithium-ion storage capability and better cyclic stability compared with unactivated counterparts. The development of high-performance rechargeable lithium ion batteries (LIBs) for efficient energy storage has become one of the components in today's information rich mobile society [114].

Microbial fuel cell (MFC) technologies are an emerging approach to wastewater treatment. MFCs are capable of recovering the potential energy present in wastewater and converting it directly into electricity. Using MFCs may help offset wastewater treatment plant operating costs and make advanced wastewater treatment more affordable for both developing and industrialized nations. In spite of the promise of MFCs, their use is limited by low power generation efficiency and high cost. Some researchers conclude that the biggest challenge for MFC power output lies in reactor design combining high surface area anodes with low ohmic resistances and low cathode potential losses. Power density limitations are typically addressed by the use of better-suited anodes, use of mediators, modification to solution chemistry or changes to the overall system design. Employing a suitable anode, however, is critical since it is the site of electron generation. An appropriately designed anode is characterized by good conductivity, high specific surface area, biocompatibility and chemical stability. Anodes currently in use are often made of carbon and/or graphite. Some of these anodes include but are not limited to: graphite plates/rods/felt, carbon fiber/cloth/foam/paper and reticulated vitreous carbon (RVC). Carbon paper, cloth and foams are among the most commonly used anodes and their use in MFCs has been widely reported employ activated carbon nanofibers (ACNF) as the novel anode material in MFC systems. Compared with other activated carbon materials, the unique features of ACNF are its mesoporous structure, excellent porous interconnectivity, and high bio available surface area for biofilm growth and electron transfer [115].

Among the diverse carbonaceous adsorbents, activated carbon fiber (ACF) is considered to be the most promisingdue to their abundant micropores, large surface area, and excellent adsorption capacity. Therefore, the investigation of formaldehyde adsorption has been steadily conducted using ACFs. However, most of precedent works have generally concerned about removal of concentrated formaldehyde in aqueous solution (formalin, and thus there was limited information whether these materials could be used in the practical application, because the concentration of formaldehyde in indoor environment was generally very low (below 1 ppm). The nitrogen containing functional groups in ACF played an important role in increasing formaldehyde adsorption ability, as also described elsewhere. However, the PAN-based ACFs still have problems in a practical application, because the adsorption capability is drastically reduced under humid condition [116, 120].

1.1.9 PORE FORMATION IN CARBON MATERIALS, PORE CHARACTERIZATION AND ANALYSIS OF PORES

About pore formation in carbon materials, this is accepted that all carbon materials, except highly oriented graphite, contain pores, because they are polycrystalline and result from thermal decomposition of organic precursors. During their pyrolysis and carbonization, a large amount of decomposition gases is formed over a wide range of temperatures, the profile of which depends strongly on the precursors. Since the gas evolution behavior from organic precursors is strongly dependent on the heating conditions, such as heating rate, pressure, etc., the pores in carbon materials are scattered over a wide range of sizes and shapes. These pores may be classified as shown in Table 1.3 [1–3].

TABLE 1.3 Classification of Pore Formation in Carbon Materials

1) Based on their origin		2) Based on their size			3) Based on their state
Intra-particle pores	Intrinsic intra-particle pores				Open pores
	Extrinsic intraparticle pores	Micropores	< 2 nm		
Inter-particle pores	Rigid interparticle pores	Mesopores	2~50 nm	Ultramicro-pores <0.7 nm	Closed pores (Latent pores)
	Flexible inter-particle pores				
		Macro-pores	> 50 nm	Supermicro-pores 0.7–2 nm	

Table 1.3 shows that based on their origin, the pores can be categorized into two classes, intra particle and inter particle pores. The intra particle pores are further classified into two, intrinsic and extrinsic intra particle pores. The former class owes its origin to the crystal structure, that in most activated carbons, large amounts of pores of various sizes in the nanometer range are formed because of the random orientation of crystallites; these are rigid inter particle pores. A classification of pores based on pore sizes was proposed by the International Union for Pure and Applied Chemistry (IUPAC). As illustrated in Table 1.3, pores are usually classified into three classes: macropores (>50 nm), mesopores (2–50 nm) and micropores (<2 nm). Micropores can be further divided into supermicropores (with a size of 0.7–2 nm) and ultra micropores (<0.7 nm in size). Since

nanotechnology attracted the attention of many scientists recently, the pore structure has been required to be controlled closely. When scientists wanted to express that they are controlling pores in the nanometer scale, some of them preferred to call the smallest pores nano-sized pores, instead of micro/mesopores [1, 2].

Pores can also be classified on the basis of their state, either open or closed. In order to identify the pores by gas adsorption (a method which has frequently been used for activated carbons), they must be exposed to the adsorbate gas. If some pores are too small to accept gas molecules they cannot be recognized as pores by the adsorbate gas molecules. These pores are called latent pores and include closed pores. Closed pores are not necessarily in small size. Pores in carbon materials have been identified by different techniques depending mostly on their sizes. Pores with nano-meter sizes, that is, micropores and mesopores, are identified by the analysis of gas adsorption isotherms, mostly of nitrogen gas at 77 K [41, 46].

The basical theories, equipments, measurement practices, analysis procedures and many results obtained by gas adsorption have been reviewed in different publications. For macropores, mercury porosimetry has been frequently applied. Identification of intrinsic pores, the interlayer space between hexagonal carbon layers in the case of carbon materials, can be carried out by X-ray diffraction (XRD). Recently, direct observation of extrinsic pores on the surface of carbon materials has been reported using microscopy techniques coupled with image processing techniques, namely scanning tunneling microscopy (STM) and atomic force microscopy (AFM) and transmission electron microscopy (TEM) for micropores and mesopores, and scanning electron microscopy (SEM) and optical microscopy for macropores [1–3].

The most important Pore Characterization methods are include: STM, AFM, TEM, Gas adsorption, Calorimetric methods, Small-angle X-ray scattering (SAXS), Small-angle Neutron Scattering (SANS), Positron Annihilation Lifetime Spectroscopy (PALS), SEM, Optical Microscopy, Mercury Porosimetry and Molecular Resolution Porosimetry. Adsorption from solution using macromolecules has been applied to macropore analysis, but we still need more examinations. It is difficult to compare one adsorption isotherm with another, but determination of the deviation from the linearity using a standard adsorption isotherm is accurate. The plot constructed with the aid of standard data is called a comparison plot. The representative comparison plots are the t and alpha plots [41, 42].

The molecular adsorption isotherm on non-porous solids, which can be well described by the BET theory. The deviation from the linearity of the t-plot gives information on the sort of pores, the average pore size, the surface area, and the pore volume. However, the t plot analysis has the limited applicability to the microporous system due to the absence of explicit monolayer adsorption. The construction of the alpha plot does not need the monolayer capacity, so that it is applicable to microporous solids. The straight line passing the origin guarantees multilayer adsorption, that is, absence of meso- and/or micropores; the deviation leads to valuable information on the pore structures. A non-porous solid has a single line passing the origin, while the line from the origin for the alpha plot of the mesoporous system bends. The slopes of the straight line through the origin and the line at the high as region give the total and mesoporous surface areas, respectively. The type of the alpha plot suggests the presence of ultra micropores and/or super micropores. Detailed analysis results will be shown for the micropore analysis [43, 46].

In the discussion of the mesopore shape, the contact angle is assumed to be zero (uniform adsorbed film formation). The lower hysteresis loop of the same adsorbate encloses at a common relative pressure depending to the stability of the adsorbed layer regardless of the different adsorbents due to the so called tensile strength effect. This tensile strength effect is not sufficiently considered for analysis of mesopore structures. The Kelvin equation provides the relationship between the pore radius and the amount of adsorption at a relative pressure. Many researchers developed a method for the calculation of the pore size distribution on the basis of the Kelvin equation with a correction term for the thickness of the multilayer adsorbed film. They so-called BJH (Barret-Joyer-Halenda) and DH (Dollimore-Heal) methods have been widely used for such calculations. As mathematical details are shown in other articles, only the simple Fortran program for the DH method. (This program can be easily used for the analysis of the mesopore size distribution).The thickness correction is done by the Dollimore-Heal equation. One can calculate the mesopore size distribution for cylindrical or slit shaped mesopores with this program. Therefore, the adsorption branch provides more reliable results. However, the adsorption branch gives a wide distribution compared to the desorption branch due to gradual uptake. Theoretical studies on these points are still done [133].

The pore size distribution from the Kelvin equation should be limited to mesopores due to the ambiguity of the meniscus in the microporous region. It is well known that the presence of micropores is essential for the adsorption of small gas molecules on activated carbons. However, when the adsorbate is polymer, dye or vitamin, only mesopores allow the adsorption of such giant molecules and can keep even bacteria. The importance of mesopores has been pointed out not only for the giant molecule adsorption, but also for the performance of new applications such as electric double layer capacitors. Thus, the design and control of mesoporosity is very desirable both for the improvement of performance of activated carbon and for the development of its new application fields [1–3].

Important parameters that greatly affect the adsorption performance of a porous carbonaceous adsorbent are porosity and pore structure. Consequently, the determination of pore size distribution (PSD) of carbon nano structures adsorbents is of particular interest. For this purpose, various methods have been proposed to study the structure of porous adsorbents. A direct but cumbersome experimental technique for the determination of PSD is to measure the saturated amount of adsorbed probe molecules, which have different dimensions. However, there is uncertainty about this method because of networking effects of some adsorbents including activated carbons and carbon nanostructures. Other experimental techniques that usually implement for characterizing the pore structure of porous materials are mercury porosimetry, XRD or SAXS, and immersion calorimetry. A large number of simple and sophisticated models have been presented to obtain a realistic estimation of PSD of porous adsorbents. Relatively simple but restricted applicable methods such as Barret, Joyner and Halenda (BJH), Dollimore and Heal (DH), Mikhail et al. (MP), Horvath and Kawazoe (HK), Jaroniec and Choma (JC), Wojsz and Rozwadowski (WR), Kruk-Jaroniec-Sayari (KJS), and Nguyen and Do (ND) were presented from 1951 to 1999 by various researchers for the prediction of PSD from the adsorption isotherms [133, 139].

For example, the BJH method, which is usually recommended, for mesoporous materials is in error even in large pores with dimension of 20 nm. The main criticism of the MP method, in addition to the uncertainty regarding the multilayer adsorption mechanism in micropores, is that we should have a judicious choice of the reference isotherm. HK model was developed for calculating micropore size distribution of slit-shaped pore; however, the HK method suffers from the idealization of the micropore filling process. Extension of this theory for cylindrical and spherical pores was made by Saito and Foley and Cheng and Ralph. By applying some modifications on the HK theory, some improved models for calculating PSD of porous adsorbents have been presented. Gauden et al. extended the Nguyen and Do method for the determination of the bimodal PSD of various carbonaceous materials from a variety of synthetic and experimental data. The pore range of applicability of this model besides other limitations of ND method is its main constraint. In 1985, Bunke and Gelbin determined the PSD of activated carbons based on liquid chromatography (LC). Choices of suitable solvent and pore range of applicability of this method are two main problems that restrict its general applicability. More sophisticated methods such as molecular dynamics (MD), Monte Carlo simulation, Grand Canonical Monte Carlo simulations (GCMS), and density functional theory (DFT) are theoretically capable of describing adsorption in the pore system. The advantages of these methods are that they can apply on wide range of pores width. But, they are relatively complicated and provide accurate PSD estimation based on just some adsorbates with specified shapes [133, 140, 145].

1.1.10 NOVEL METHODS FOR CONTROLLING OF PORE STRUCTURE IN CARBON MATERIALS

1.1.10.1 ACTIVATION

Activation processes are often classified into two, gas activation and chemical activation. Different oxidizing gases, such as air, CO_2, and water vapor, were used for gas activation. For chemical activation, $ZnCl_2$ and KOH were used as an activation agent. Recently, a great success to get a high surface area reaching to approximately 3600 m² g⁻¹ was obtained by using KOH for activation and applied for the preparation of some of activated carbons. During activation process, the creation of micropores is the most important. In most of carbon materials, however, macropores and mesopores coexisted with micropores. In other words, macropores and mesopores had to be formed during the activation process in order to develop a large number of micropores. On adsorption and desorption procedure, however, these pores play an important role as diffusion pathways for adsorbates to micropores. For the carbons prepared from biomasses, such as coconut shells and wood chips, many macropores are already formed during carbonization as a memory of cell structure of the original biomasses, which seems to make micropore development by activation easier. In activated carbon fibers, however, micropores are formed on the surface of thin carbon fibers. Such a direct exposure of micropores to adsorbates gives an advantage of fast adsorption/desorption. The pore development in carbon materials in nanometric scale by air oxidation was studied in detail, which is an activation process with the simplest in the equipments, the mildest in thermal conditions, and also energy and resources saving among different activation processes. The commercially available carbon spheres having the diameter of approximately 15 μm were activated at different temperatures of 355–430°C for various residence times of 1–100 h in a flow of dry air.

The original carbon spheres had negligibly small BET surface area, no pores on their surface were observed with high magnification scanning electron microscopy, and their structure was amorphous with non-graphitizing random nanotexture, so called glasslike carbon. Therefore, oxidation is supposed to start only on the physical surface of each sphere and proceed to the inside of the spheres by forming macropores, mesopores, and micropores. Each experimental point on oxidation yield against logarithm of residence time, log t, at different temperatures are superimposed on the curve for a reference temperature of 400°C by the translation along the log t axis to give a smooth curve, being called the master curve. Plot of shift factors against inverse of oxidation temperature can be approximated to be linear and gives apparent activation energy AE of approximately 150 kJ mol⁻¹ from its slope. In wet air, however, AE of approximately 200 kJ mol⁻¹ was obtained at the same temperature range. For different pore parameters measured by BET, as BJH, and DFT methods, the master curves could be obtained by applying the same shift factors. Master curves for micropore volume determined by as plot, ultramicropore and supermicropore volumes by DFT method and mesopore volume by BJH method are shown in order to make the comparison easier, together with some SEM images to show the appearance of the spheres. In the beginning of oxidation, up to 10 h oxidation at 400°C, the main process is the formation of ultramicropores. Above 10 h up to approximately 60 h, relative amount of ultramicropores decreases but supermicropores increase with increasing oxidation time. Above 65 h, both ultramicropores and supermicropores decrease rapidly but mesopores increases slightly, which result in the decrease in surface areas. This change in pore volumes might suggest the gradual enlargement of pore size from ultramicropore through supermicropore to mesopore and possibly to macropores by prolonged oxidation above 100 h. At very beginning of activation, 1 h oxidation at 400°C, BET surface area reaches 400 m^2/g of which the predominant part is microporous surface area. This was experimentally proved to be caused mainly by the change of closed pores, which were formed during carbonization of the precursor phenol resin, to open pores. To understand the activation process from the view point of gasification, it was proposed to normalize the fractional weight loss in different atmospheres (different oxidizing agents, such as steam and CO_2, and their different pressures) as a function of t/t_o 5, where to is the time giving fractional weight loss of 0.5. The experimental data of mass loss obtained at a constant temperature for each oxidizing agent were successfully unified to one curve. The activation process of glass-like carbon spheres in air at different temperatures and residence times was shown to be understood by master curves for the yield and pore structure parameters. On the same carbon spheres, adsorption behaviors of various organics in their aqueous solutions were also understood by the master curves for each adsorbates as functions of oxidation temperature and time. Unification curves are shown as a function of dimensionless time t/t_o 5 and master curves are expressed by real time at a reference temperature. The former seems to be useful to compare the activation (gasification) of various carbonaceous materials and to discuss its mechanism, but the latter might be useful to discuss the activation conditions to prepare activated carbons.

The derivation procedure of master curves suggests that the conversion between oxidation temperature and time was possible for pore structure parameters as well as activation yield through air oxidation. Activation processes have been pointed out to have some demerits. The mesopores can usually be created by the enlargement of micropores. In other words, by some expense of micropores, and certain part of carbon atoms has to be gasified to CO and/or CO_2 during activation process, in other words, the final yield of activated carbons becomes low. These demerits were pointed out to be one of the barriers for cost down of the industrial production of activated carbons. Also these demerits of activation process were one of the motive forces for the development of new preparation processes of porous carbons, as described in the following sections. Activation in two steps in air was reported to be efficient, the first step at a high temperature for a short time, followed by the second step at a low temperature for a long residence time. On glass like carbon spheres, two step activation at 500°C for 3 h for the first step and at a temperature below 415°C for different periods for the second step. In the beginning of activation, activation yields larger than 60 wt.%,higher S_{BET} and S_{mw} are obtained with the same yield, in other words, the same surface areas can be obtained with about 10 wt.% higher yield by two step activation than by one step activation.

1.1.10.2 TEMPLATE METHODS

Microporous carbons, of which the highest surface area and pore volume were approximately 4000 g^{-1} and 1.8 mL/g, respectively, were prepared through the carbonization of a carbon precursor in nano channels of zeolite, with the procedure called as template carbonization technique. Since the size and shape of the channels in zeolites are strictly defined by their crystal structures and the pores formed in the resultant carbon are inherited from their channels, micropores formed in the resultant carbon are homogeneous in both size and morphology. Impregnation of furufuly alcohol (FA) into the channels of a zeolite was carried out under vacuum at a low temperature, followed by washing excess FA attached on physical surface of zeolite particles with mesitylene. The composite particles of FA/zeolite thus obtained were heated at 150°C for 8 h to polymerize FAs (PFA/zeolite composite). The composites were carbonized at 700°C, followed by dissolution of the template zeolite by 46–48% HF solution. The relation of zeolite cage to the resultant pores in carbon. Detailed preparation procedure for these highly microporous carbons was reviewed. High resolution TEM images show that regular alignment of super cages with the size of 1.4 nm is inherited in the resultant carbon as the periodicity of approximately 1.3 nm. The carbon gives a diffraction peak at an angle of approximately 6° in 2θ of CuKa X-rays, just as the template zeolite does, which corresponds to the periodicity of approximately 1.3–1.4 nm due to superstructure. Very high surface area indicates the presence of the curved carbon surfaces, which was predicted by grand canonical Monte Carlo simulation. Mesoporous carbons were also prepared by template method using mesoporous silica. By coupling with an activation process, micropores could be easily introduced into these mesoporous carbons [21, 40].

A simple heat treatment of thermoplastic precursors, such as poly(vinyl alcohol) PVA, hydroxyl propyl cellulose HPC, polyethylene terephthalate) PET, and pitch in the coexistence of various ceramics, was able to coat ceramic particles by porous carbon. Using MgO particles as substrate ceramics, it was experimentally proved that the carbons formed from PVA at 900°C were experimentally shown to have larger surface area after the substrate MgO was dissolved out by a diluted acid to isolate carbon. The carbons have high S_{BET}, particularly the carbons obtained from the mixtures prepared from Mg acetate and citrate with PVA in their aqueous solutions (solution mixing). Most of carbons show a relatively high external surface area Sext, which is known to be mostly due to mesopores. Mesopores formed in the carbon was known to have almost the same size as MgO particles, which were formed by the thermal decomposition of Mg compounds in advance of the thermal decomposition and carbonization of carbon precursor. When MgO particles with rectangular morphology were used, the rectangular pores with the same size were observed under SEM in the resultant carbons. Therefore, it was concluded that MgO works as a template for mesopore formation. Pore size distributions in micropore and mesopore ranges measured by DFT and BJH methods, respectively. The carbons formed on the surface of MgO particles are microporous, rich in the pores with the width of approximately 1 nm. The size of mesopores formed in the carbon is found to depend on the starting MgO precursors, Mg acetate mixed with PVA in solution forming mesopores with the size of approximately 10 nm, Mg citrate those with 5 nm, and Mg gluconate those with 2–4 nm. In the case of Mg acetate/PVA mixtures, the sizes of mesopores depend strongly on mixing method mixing in powder (powder mixing) gives a broad distribution of pore size but solution mixing gives a relatively sharp distribution of mesopores at around 10 nm. However, Mg citrate/PVA mixtures gave almost the same size distribution of mesopores by mixing in either powder or solution. The substrate MgO had an additional advantage that it could be easily dissolved out at room temperature by a diluted solution of acid, even by 1 mol/L citric acid, so that MgO was experimentally proved to be recycled. The preparation of mesoporous carbons using MgO template and their applications were reviewed. Porous carbons containing both micropores and mesopores were also prepared by using Ni hydroxide template. Aqueous suspension of Ni $(OH)_2$, which was prepared from $Ni(NO_3)_2$ and NaOH, was mixed with ethanol solution of phenolic resin and then carbonized at 600°C after drying. By dissolving inorganic species, formed during the process (NiO and Na_2CO_3), porous carbon was isolated. The carbon obtained had S_{BET} of 970 m^2/g, total pore volume of 0.69 mL/g, and micropore volume of 0.3 mL/g, in which the predominant micropore and mesopore sizes were approximately 0.8 nm and 15 nm, respectively [26, 27].

1.1.10.3 DEFLUORINATION OF PTFE

Porous carbons were reported to be prepared through defluorination of poly tetrafluoroethylene (PTFE) with alkali amalgamates. The detailed studies were reported for the preparation of porous carbons from PTFE by using different alkali metals. PTFE film was pressed with lithium metal foil under 4 MPa in Ar atmosphere for 48 h to defluorinate PTFE. After excess lithium metals were washed out with methanol, the heat treatment at 700°C and washing by dilute HC1 were carried out in order to eliminate finely dispersed LiF. Defluorination of PTFE was also possible through heating a mixture of PTFE powders with alkali metals, Na, K, and Rb, in vacuum at 200°C in a closed vessel. N_2 adsorption isotherms of resultant carbons were found to depend strongly on alkali metal used. Defluorination of PTFE with Na metal was found to give mesopore rich carbon and very high S_{BET} as 2225 m^2/g. S_{BET} of these carbons prepared using Na was found to increase with heat treatment at a high temperature up to 1000°C, probably because of gasification of carbon with surface oxygen functional groups. Deflourination of PTFE using Na metal has an advantage, Na metal is much cheaper in price and easier to handle, and so the process being simpler than using other alkali metals.

The irradiation of PTFE before carbonization is also preferable for getting high surface area. Defluorination of PTFE was possible in 1, 2-di-methoxyethane

(DME) solution of alkali metals naphthalene complexes at room temperature. S_{BET} of the resultant carbon was 1000–1800 m ^{2}/g and no effect of alkali metals on pore structure was observed.

1.1.10.4 CARBON AEROGELS

Carbon aerogels, which have been well known as one of mesoporous carbons, were prepared from the pyrolysis of organic aerogels of resorcinol and formaldehyde. Extensive studies focused on their pore structure and also on doping of some metals were carried out. Primary carbon particles have the size of approximately 4–9 nm and interconnected with each other to forma network. Adsorption isotherms belong to type IV and have a clear hysteresis. Pore structure parameters calculated through a analysis are listed up on carbon aerogels prepared from resorcinol and formaldehyde at different temperatures can reported. These carbon aerogels contain predominantly inter particle mesopores formed in a three-dimensional network of interconnected minute carbon particles, and only small amount of intra particle micropores were formed in primary carbon particles. Carbon aerogels could be activated in order to increase micropores by CO_2 at 900°C. The activation for 5 h increased both of micropores and mesopores: pore volume of 0.68 and 2. 04 mL g and surface area of 1750 and 510 m^2/g, respectively. The detailed studies on adsorption of N_2 at 77 K and of water vapor at 303 K on activated carbon aerogels, whose surface functional groups were showed clearly that the amount of adsorbed water corresponded only to the micropore volume and not to the mesopore volume. Doping of Ce and Zr into carbon aerogels was found to result in micropore rich carbon materials. Pore structure of carbon aerogels was known to be governed by that of precursor organic aerogels, which was controlled by the mole ratios of resorcinol to formaldehyde (R/F), to water (R/W) and also to basic catalyst Na_2CO_3(R/C). Aqueous gels synthesized were dried under supercritical CO_2 Pore size distributions in mesopore region for both original organic aerogels and resultant carbon aerogels as a function of R/W, the other factors R/F and R/C being the constant as 0.5 and 75, are reported. Pore size distributions of organic aerogels are rather sharp and their maxima decrease with increasing R/W ratio. By carbonization of these organic aerogels, pore size distributions shift to a little smaller size, mainly because of the shrinkage of gels during thermal decomposition. Instead of supercritical drying of aqueous gels, freeze-drying method was also applied. On the gels prepared through freeze drying (cryogels), much smaller shrinkage in pore size during carbonization was observed. Conditions for the preparation of resorcinol-formaldehyde gels through solgel condensation and those for freeze-drying were studied in detail in order to control mesoporosity in resultant carbons. Effect of drying process of gels, such as freeze drying, microwave drying, and hot air drying, on pore structure were also studied. The first two drying methods are effective in order to get mesoporous carbons. Resorcinol-acetaldehyde cryogels could be also a precursor for mesoporous carbons. Pore structure in the resultant carbons was found to depend on pH- value in the solution, which was changed by mixing ratio of R/C. Pore volumes of the resultant carbon aerogels after 800°C carbonization with pH of 8.0 (R/C = 25), no porous carbon is obtained. Porous carbons are formed in the pH range of 7.0–8.0, particularly mesopore-rich in the pH range of 7.3–7.7. Carbon aerogels were used as a template for the preparation of highly crystalline zeolite with uniform mesoporous channels [1, 2, 39].

1.1.10.5 POLYMER BLEND METHOD
Polymer blend method was proposed to synthesize various types of carbons, mixing two different polymers, one having a high carbon yield, such as polyfurufuryl alcohol, and the other, a low carbon yield, such as polyethylene. The scheme of polymer blend method to get carbon balloons, carbon beads, and also porous carbons is shown in Fig. 1.2. By applying spinning on blended polymer, certain success in obtaining carbon nanofibers was reported. Through the synthesis of polyurethane-imide films and their carbonization, carbon films were obtained, of which macropore structure was controlled by changing the molecular structure of polyurethane. Prepolymer poly(urethane-imide) films were prepared by blending polyamide acid giving polyimide (PI) with polyurethane (PU). Polyurethane-imide films after heating up to 200°C showed phase separation of PI and PU, where the former polymer formed a matrix and the latter formed small islands. By heat treatment up to 400°C, PU component was pyrolyzed to gases and resulted in porous PI films, which can be converted easily to porous carbon films by carbonization. Pore sizes in these carbon films were controlled by the blending ratio of PI to PU and also by the molecular structure of PU [1, 2].

1.1.10.6 SELECTION OF SPECIFIC PRECURSORS

1.1.10.6.1 DERIVATIVES OF BUTADIYNYLENE
Poly(phenylene butadiynylene) were found to give very high carbon yield of more than 90 wt.% after the heat treatment at 900°C, very close to theoretical yield, whose molecular structure are determined. The resultant carbons are amorphous state and microporous, having total surface area of 1 330 rn^2/g microporous surface area of 1 300 rn^2/g and micropore volume of 0 49 mLg^{-1}. Their pyrolysis behavior was characterized by a very sharp exothermic peak at around 200°C without accompanying any mass change, which was due to 1, 4-polymerization of butadiynylene moiety and resulted in cross-linking between molecules, and a little but gradual mass loss of approximately 600°C. The material heat-treated above 200°C was so highly and strongly cross linked that the hydrogen atoms that remained were mostly stripped off as hydrogen molecules, which is the main reason to give high carbon yield. The derivatives with methyl radicals on benzene ring gave also high carbon yields, close to 80 wt.% of carbon content [1].

1.1.10.6.2 FLUORINATED POLYAMIDES

Microporous carbon films were prepared from aromatic polyamides synthesized from different dianhydrides and diamines with different molecular structures by carbonization. Structures of repeating unit in polyimides are reported. They were characterized by the presence of pending groups, $-CH$, and $-CF$, and number of phenyl rings in the repeating unit. Highly microporous carbon films were obtained without any activation process, of which microporous surface area was closely related to the contents of phenyl ring [1, 2].

1.1.10.6.3 METAL CARBIDES

Various metal carbides were found to give highly microporous carbon through the heat treatment in a flow of carbide-derived carbons. S_{BET} was reported to be 1000–2000 m^2/g^{-1}, depending on the precursor carbide and heat treatment temperature. The carbide-derived carbons gave a relatively sharp distribution in pore size and their predominant pore size depends mostly on the precursor carbide; SiC, TiC and Al_4C_3 gave 0.7–0.8 nm, and B_4C approximately 1.3 nm and Ti_3SiC_2 0.5–0.6 nm. From ZrC_{098}, however, the carbons were obtained with a sharp size distribution of approximately 0.7 nm after the heat treatment at 300–600°C, but with a broad distribution ranging from 0.6 to 3 nm after the heat treatment at 800–1200°C. These carbide-derived microporous carbons were used as the electrode of electric double layer capacitors.

1.1.10.6.4 BIOMASSES

The cores of kenaf plant (Hibiscus cannabinus) were found to give a high S_{BET} as high as 2700 m^2/g by the heat treatment, without any additional activation process. This high surface area was supposed to be due to the departure of metallic impurities (mostly K), which were originally included in the cores, during carbonization. Porous carbons were successfully prepared from chips of cypress (Cupressus) by the heat treatment under a flow of superheated steam. Pore structure in cypress charcoals could be controlled by changing carbonization conditions (temperature of superheated steam, and supplying and transferring rates of chips).

1.1.10.7 SELECTION OF PREPARATION CONDITIONS

1.1.10.7.1 MOLECULAR SIEVING CARBON FILMS

Carbon membranes with molecular sieving performance were successfully prepared by the carbonization of commercially available polyimide films with the thickness of 0.1 mm. adsorbed amounts of gas molecules with different sizes are shown on the carbon films heat treated at different temperatures, ethane C_2H_6 with the size of 0.40 nm permeating in a large amount through the film heated up to 700°C, but with very small amount through the one heated up to 1000°C. This result suggested a strong dependence of pore sizes in carbon films on carbonization temperature, which was supposed to be due to shrinkage during carbonization [1, 2, 54, 55].

1.1.10.7.2   MACROPOROUS CARBONS (CARBON FOAMS)

Carbon foams were prepared from mesophase pitch through either blowing or pressure release of molten pitches followed by stabilization in air. Because the foam with a high thermal conductivity attracted attention for a potential filler material of some composites, the preparation of a large sized carbon foam with high thermal conductivity from mesophase pitch by a new and less time consuming process was extensively studied. The graphitized foams have the bulk density of 0.2–0.6 gcm^{-3}, with an average pore size of either 275–350 micrometers or 60–90 micrometers and relatively high graphitization degree.

1.1.10.7.3   INTERCALATION

Intercalation of various species into graphite gallery can be considered to be the chemical adsorption into intrinsic pores, which are flexible and two-dimensional slit-shaped micropores. These intercalation phenomena are one of characteristics of carbon materials in graphite family. In addition, intercalation can be understood as a new process to create extrinsic pores in the compounds, of which the size is controllable by the size of intercalates. In graphite intercalation compounds (GICs) of alkali metals with 2D structure, for example, nanospaces are formed surrounded by alkali metal ions and graphite layer planes. As an example, arrangement of Cs$^+$ ions in graphite gallery. In this compound, the space with the size of approximately 0.311 × 0.266 nm can accept the third component. Changing the alkali metal ions to K (ion radius of 0.133 nm) and Rb (0.148 nm) from Cs$^+$(0.169 nm), the size of the space for the third component changes [1, 2].

1.1.10.7.4   EXFOLIATION

Exfoliated graphite has been produced as a raw material for flexible graphite sheets in a huge amount in the world. Usually it is produced through rapid heating of residue compounds of natural graphite flakes with sulfuric acid up to a high temperature as 1000°C. It consists of fragile worm-like particles, which are formed by exfoliating preferentially along the normal to the basal plane of graphite. There are at least three kinds of pores in exfoliated graphite. Large pores are formed mainly by the complicated entanglement of worm-like particles during abrupt exfoliation of each graphite flakes and so they are flexible interparticle macropores. In addition, crevice-like pores on the surface of worm-like particles and pores inside the particles are formed,respectively. In order to evaluate the pores, which are rigid intraparticle, pores, a technique to prepare fractured cross-section of worm-like particle had to be developed and various pore structure parameters were determined with the aid of image processing technique.

1.1.11   IMPORTANCE AND NECESSITY OF CONTROLLING OF PORE STRUCTURE IN CARBON MATERIALS

1.1.11.1   IRREVERSIBLE ADSORPTION OF CO_2 GAS

A glass-like carbon sphere was found to show an irreversible adsorption of CO_2 gas. Two commercially available glass-like carbon spheres APS and APT were used, which were prepared at 700°C and 1000°C, respectively, in CO_2 atmosphere. The size of the spheres was rather homogeneous, at approximately 15 micrometers. The adsorption/desorption behavior of CO_2 was measured volumetrically at different temperatures and gravimetrically at 0°C. When CO_2 pressure in the electro balance was less than 1 atm, the adsorbed amount of CO_2 at the saturation on APT decreased, although the behavior during adsorption, evacuation, and heating was the same. Selective adsorption of CO_2 was shown to be possible by the carbon spheres after the irradiation of oxygen plasma. These gravimetric studies on the adsorption/desorption of CO_2 suggest the presence of three types of micropores in APT, depending on the accessibility of CO_2. The first type of micropores adsorbs CO_2 molecules immediately and also desorbs quickly only by evacuation at 0°C, of which the volume is supposed to be approximately 40%. The second type of micropores adsorbs CO_2 very slowly, taking approximately 200 h to reach saturation and requires heating to 250°C under vacuum for desorption, being about 55%. From the third type of micropores, the adsorbed CO_2 is not released even by heating up to 250°C under vacuum (strongly trapped CO_2 about 5% in volume). The detailed study by supercritical gas adsorption analysis with the aid of grand canonical Monte Carlo simulation showed that the width of entrance (mouth) of ultramicropores in APT is a little narrower than that of inside of pores [1, 2].

1.1.11.2   CARBON FOAM FOR WATER VAPOR ADSORPTION/DESORPTION

In order for microporous carbons to play certain functionality, micropores are preferred to be directly open to adsorbates in most applications for their easy and rapid adsorption/desorption. To keep a large number of micropores open, morphology of carbon materials is important, as mentioned on activated carbon fibers. Carbon foams can also satisfy this requirement. Carbon foams were prepared from a fluorinated polyimide (6FDA/TFMB), which was known to give microporous carbon, by using either urethane or melamine foam as a template and by sub sequent activation in air at 400°C. Their adsorption/ desorption behaviors for water vapor were examined in TG apparatus. Many points are remained to be studied; optimization of pore structure for adsorption/desorption of water vapor with higher rate and also at lower relative pressure, search for more appropriate template foam and impregnant polymer, etc.

1.1.11.3   POROUS CARBON FOR CAR CANISTER

A strong demand to recover gasoline vapor during parking of cars is now understood to be important for saving gasoline and also for avoiding the contamination of air. Gasoline vapor is absorbed into activated carbon in a device called "canister" during parking and reused by passing air through adsorbed activated carbon during running. For activated carbon in the car canister, a specific pore structure is required. For the evaluation of the performance of car canister, adsorption/desorption of butane gas has often been used. In the recent work, direct gravimetric measurement of adsorption of gasoline vapor was carried out, which was not a standardized method for car canister, and also no experiment on desorption of gasoline vapor was done. The confirmation of the results is demanded by using butane gas [1].

1.1.11.4   CARBON ELECTRODE FOR ELECTRIC DOUBLE LAYER CAPACITORS (EDLC)

As mention before in Section 1.4,because EDLC is on the basis of the formation of electric double layer at the electrode/electrolyte interface, which is due to the physical adsorption of electrolyte ions, activated carbons with high surface area are usually used as electrode materials and their pore size distribution (PSD) is pointed out. To have an influence in EDLC performance Capacitance of EDLCs was possible to be explained by dividing into two capacitances of the surface due to micropores (microporous surface area) and of that due to larger pores (evaluated as external surface area) on different activated carbons. Porous carbons are characterized by extremely large BET surface areas that range from 500 to 3000 $m^2 g^{-1}$. This surface area largely arises from a complex interconnected network of internal pores. Micropores have a high surface area to volume ratio and, consequently, when present in significant proportions are a major contributor to the measured area of high surface area activated carbons. Micropore sizes extend down to molecular dimensions and play an important role in the selectivity of adsorption-based processes, through restricted diffusion and molecular sieve effects. Fine micropores also exhibit a greater adsorbent adsorbate affinity due to the overlap of adsorption forces from opposing pore walls. Accordingly, adsorption in fine pores can occur via a pore filling mechanism rather than solely by surface coverage (as is assumed by the Langmuir and BET calculations of surface area). In such cases, the conversion of adsorption data into an estimate of surface area, by the application of the BET equation, can lead to unrealistically high surface area estimates. Clearly, the pore size distribution of porous carbons influences to a large degree the fundamental performance criteria of carbon-based super-capacitors, the relationship between power and energy density, and the dependence of performance on frequency [20, 34, 45].

Not surprisingly, therefore, considerable research is presently being directed towards the development of carbon materials with a tailored pore size distribution to yield high capacitance and low resistance. Electrodes Electric double layer capacitors (EDLCs) are expected to be one of the promising devices for electric energy storage and a variety of applications have been developed, for example,for memory back-up, cold start vehicle assist, storage for solar cell power, and also for high power sources, such as power trains and electric vehicles. Extensive research works are still carried out in order to obtain better performance. Asymmetric EDLCs using different activated carbons in two electrodes, negative and positive electrodes, were constructed and their performance was studied in non-aqueous electrolyte. The capacitance and rate performance of these asymmetric EDLCs were found to be governed predominantly by the pore structure of the carbon in the negative electrode [108].

1.1.11.4.1 SUPER-CAPACITORS BASED ON ACTIVATED CARBON MATERIAL

More than 20 years ago an experimental super-capacitor cell by using commercial activated carbon fiber (ACF) cloth for each of the two electrodes and glass fiber filter paper as separator in organic electrolyte was realized. At that time 36.2Wh/kg specific energy, 11.1 kW/kg specific power and 36.5 F/g specific capacitance were estimated. The specific capacitance was considered per gram of ACF. For the ACF cloth made from phenolic resin, a specific surface area of 1500–2500 m^2/g was estimated. The specific energy and specific power reached in practice at this time are much lower than the above estimated values because these take into consideration the overall weight of capacitor cell including its package. Activated carbon composite electrodes for electrochemical capacitors have been also investigated. Thus for hydrous ruthenium oxide/activated carbon electrode in H_2SO_4 electrolyte, an increase of specific capacitance from 243 F/g (for pure activated carbon electrode) to 350 F/g for composite electrode where 35% is ruthenium oxide is reported by scientists.

In other work for only 3.2% ruthenium oxide in the composite electrode, an increase in the capacitance of 25% to a value of 324 F/g is reported. Other experiments with ruthenium oxide/activated carbon composites used at positive electrodes in electrochemical capacitors indicated increase of the specific capacitance. Nickel hydroxide/activated carbon composite electrodes used in electrochemical capacitors provide significant increase of specific capacitance from 255 to 314 F/g. If manganese oxide/activated carbon composite electrodes are used increase of specific capacitance takes place. Other materials used for activated carbon composite electrodes have been found to increase the specific capacitance. In hybrid or asymmetric electrochemical capacitors, one electrode is based on activated carbon material and another one is based on another material (nickel hydroxide, manganese oxide, etc.). Higher specific capacitance or specific energy is possible than in the case of symmetric capacitors, based only on activated carbon electrodes [34, 45, 64].

It is usually anticipated that the capacitance of a porous carbon (expressed in F g^{-1}) will be proportional to its available surface area (in m^2 g^{-1}). Whilst this relationship is sometimes observed, in practice it usually represents an oversimplification. The major factors that contribute to what is often a complex (nonlinear) relationship are:(i) assumptions in the measurement of electrode surface-area; (ii) variations in the specific capacitance of carbons with differing morphology; (iii) variations in surface chemistry (wettability and pseudo capacitive contributions), (iv) variations in the conditions under which carbon capacitance is measured. The surface areas of porous carbons and electrodes are most commonly measured by gas adsorption (usually nitrogen at 77 K) and use BET theory to convert adsorption data into an estimate of apparent surface area. Despite its widespread use, the application of this approach to highly porous (particularly microporous) and heterogeneous materials has some limitations and is perhaps more appropriately used as a semi-quantitative tool. Possibly the greatest constraint in attempting to correlate capacitance with BET surface area, is the assumption that the surface area accessed by nitrogen gas is similar to the surface accessed by the electrolyte during the measurement of capacitance. While gas adsorption can be expected to penetrate the majority of open pores down to a size that approaches the molecular size of the adsorbate, electrolyte accessibility will be more sensitive to variations in carbon structure and surface properties. Electrolyte penetration into fine pores, particularly by larger organic electrolytes, is expected to be more restricted (due to ion sieving effects) and vary considerably with the electrolyte used. Variations in electrolyte–electrode surface interactions that arise from differing electrolyte properties (viscosity, dielectric constant, dipole moment) will also influence wettability and, hence, electrolyte penetration into pores [64, 68].

1.1.11.4.2   SUPER-CAPACITORS BASED ON CNTS

The presence of mesopores in electrodes based on CNTs, due to the central canal and entanglement enables easy access of ions from electrolyte. For electrodes built from multi-walled carbon nanotubes (MWCNTs), specific capacitance in a range of 4–135 F/g was found in Refs. For single-walled carbon nanotubes (SW-CNTs) a maximum specific capacitance of 180 F/g and a measured power density of 20 kW/kg at the energy density of 7Wh/ kg, in KOH electrolyte is reported. In other work an initial specific capacitance of 128 F/g decreased after charging–discharging cycles to 58 F/g. Enhancement of specific capacitance given by CNTs is possible For example, by their mixing with conducting polypyr role. A comparative investigation of the specific capacitance achieved with CNTs and activated carbon material reveals the fact activated carbon material exhibited significantly higher capacitance. SWCNTs/polypyr role nanocomposite electrode used in recent work indicated much higher specific capacitance than pure polypyr role or SWCNTs electrodes. Super capacitor electrodes based on carbon nanotube–polyaniline (CNT–PANI) nano composite by coating polyaniline on the surface of the CNT have been used recently. At a current density of 10 mA/cm, the CNT–PANI nano composite exhibits high specific capacitance of 201 F/g, in comparison with a value of 52 F/g for the CNT. The super capacitors based on the CNT–PANI nano composite have an energy density of 6.97 Wh/kg and an outstanding power performance [65].

1.1.11.5 MACROPOROUS CARBON FOR HEAVY OIL SORPTION AND RECOVERY

Exfoliated graphite was found to sorb a large amount of heavy oils at room temperature very quickly, more than 80 kg.kg^{-1} within 1 min. Different carbon materials were studied through the determination of sorption capacity, sorption kinetics, and repeated sorption/desorption cycles in order to recover spilled heavy oils and also to recycle both heavy oils and carbon materials. Exfoliated graphite was then applied to sorb other oils, such as cooking and engine oils and also to organics with large molecules, such as biofluids. Exfoliated graphite with a low bulk density has very high sorption capacity for heavy oil, but its sorption rate is rather low. By increasing its bulk density, the sorption rate can be improved slightly, but sorption capacity decreases quickly at the same time. Carbonized fir fibers have a sorption capacity comparable with the exfoliated graphite and the same dependence on bulk density. By densification of carbonized fiber felts, however, sorption rate increases rather rapidly. These high sorption capacities for the carbon materials were found to be mainly because of flexible macropores in these carbon materials. Macropores with the size in the range of 1–600 micrometers are primarily responsible for heavy oil sorption [1].

1.1.12 RECENT STUDY WORKS ABOUT CONTROLLING OF PORE SIZE

Summary of some recently reported papers about confirmal modeling of controlling of pore size in carbon based nano adsorbent are presented in Table 1.4 [8, 40, 121]:

TABLE 1.4 Summary of Recently Confirmal Models for Controlling of Pore Size in Carbon Based Nano Adsorbents

Carbon Material Type	Applied Model and Simulation Methods	References
AC, ACF, MSC	N$_2$ adsorption at77 K and t-plot, Alpha-plot	[8]
Carbon structures	N$_2$ adsorption at77 K and Alpha-plot	[9]
AC	G CMC and DFT (Monte carlo simulations)	[10]
Carbon structures	N$_2$ and NLDFT model	[11]
AC	G CMC	[12]
ACF	H$_2$S adsorption	[13]

TABLE 1.4   *(Contiued)*

Carbon Material Type	Applied Model and Simulation Methods	References
C-ZIFs	N_2 adsorption at77 K and t-plot, BET	[14]
Carbon structures	ND and DFT	[15]
AC	N_2 adsorption at77 K and t-plot, Alpha-plot	[16]
AC	Monte Carlo simulations	[17]
p-carbon	NLDFT- BJH method	[18]
ACF and AAPFs	DR equation	[19]
AAC	N_2 adsorption at77 K and t-plot, Alpha-plot	[20]
ACH structure	N_2 adsorption at77 K and t-plot, Alpha-plot	[21]
ACF	N_2 adsorption at77 K and t-plot, Alpha-plot	[22]
AC	Ar adsorption at87 K and t-plot	[23]
ACF	Ar adsorption at87 K and Alpha-plot	[24]
AC	IAST-Freundlich model	[25]
ACF	Ar adsorption at87 K and t-plot, BET	[26]
MSC,AC	N_2 adsorption at77 K and t-plot, Alpha-plot	[27]
P-ACS	N_2 adsorption at77 K and t-plot, BET	[28]
AC	GCMS and SIE equation – methane adsorption	[29]
AC	N_2 adsorption at77 K and t-plot, Alpha-plot	[30]
Carbon structures	Ar adsorption at87 K and t-plot,Alpha-plot	[31]
Carbon structures	N_2 adsorption at77 K and t-plot, Alpha-plot	[32]
Carbide-derived carbons	Ar adsorption at87 K and t-plot	[33]
C-xerogels	N_2 adsorption at77 K and Alpha-plot, BET	[34]
Carbide-derived carbons CDC's	N_2 adsorption at77 K and t-plot, Alpha-plot	[35]
Carbon black	NLDFT	[36]

TABLE 1.4 *(Contiued)*

Carbon Material Type	Applied Model and Simulation Methods	References
Carbon Structures	N_2 adsorption at77K and t-plot, Alpha-plot	[37]
Glassy carbon	Monte carlo simulation N adsorption	[38]
Carbon structures	DFT – Ar adsorption at 77 K	[39]
AC	N_2 adsorption- and DR equation – BjH	[40]
P-ASC	N_2 adsorption at 77 K – BjH method	[123]
ACF	DRS equation–N_2 adsorption at 150 ^{c}	[126]
AC	DFT – N_2 adsorption at 77 K	[128]
CNT/Polymer Composite	MDS – NPT	[132]
CNT composite	Molecular Dynamics Simulation (MDS) – PMF	[121]
SWCNT	GCMC-LJ potential	[131]
CNT	MDS – PEOE algorithm–PME	[124]
CNT/PE composite	MDS – Brenner – Newtonian equation	[125]
SWCNT	MDS – DFTB-CPMD-REBO	[127]
CNT	MDS	[122]
CNT/Polymer	MDS – SHAKE algorithm – DL – Poly	[130]
SWCNT	MDS – CPMD-DFT	[129]
AC	DS-HK-IHK	[133]
2D Graphene	ADS – DFT – APMD – PAW	[134]
SWCNT	MDS – Berenner	[135]
CNT ropes	GCMC	[136]
CNT/Sodium	MDS	[137]
CNT	Abintio QMS-SFC	[138]
CNT	MDS – USHER algorithm – LJ Potential	[139]

TABLE 1.4   *(Contiued)*

Carbon Material Type	Applied Model and Simulation Methods	References
Nanotube	MDS – PES – Verlet algorithm	[140]
CNF	MDS – GRASP – RFF	[141]
SWCNT	MDS – DFT – B3LYP	[142]
CNT	MDS – LJ potential	[143]
Metal Membranes	EMD – GCMC – LJ potential	[144]
CNT	MDS–LJ – TIP5P	[145]

Recent research works about controlling of pore structure in carbon materials that categorized as fallowing include: First, using the grand canonical Monte Carlo (GCMC) method, or other simulation methods to determine adsorption isotherms in Ar (87 K) or N_2 (77 K) that are simulated for all the carbon sample structures to reach optimum condition in experimental works. Second, experimental works to obtained PSD by t-plot or alpha plot curves and BET equation that determine other adsorption parameters and show that samples structures maybe micro or mesoporous (with different ratio of micro/mesopores). Finally PSD are calculated using the Horvath–Kawazoe (HK), density functional theory (DFT), D-R method, Barrett–Joyner–Halenda (BJH) approaches and other mathematical methods and this model results with those predicted by the experimental work results compared and adapted, to prove selected mathematical model is significant and simulation that applied is verified [121].

This review is concerned with such pore controlling methods and models performed by researchers. An ultimate aim of these researchers is to establish a model, which can tailor pore structure of carbon materials to read any kind of requirement. In this review, we would like to highlight how effort the researchers have made to control micro and mesopores in carbon materials, and how active they are in achieving the final target and prepare them for special application that here our means adsorption contaminates from agues environments. This is one of the first studies in which different methods of calculation of PSDs for carbon nano structures from adsorption data can be really verified, since absolute PSDs are obtained using the certain method. This is also one of the first studies reporting the results of computer simulations of adsorption on carbon structure models. There are well documented reports in the area of PSD estimation and pore structure control modeling from adsorption data and compare of this results with theoretical predictions; for example, effect of chemical ratio and compare of some type of activating agent on pore structure of carbon materials, with some different classical models, but it appears there is still a lack of studies on the modeling and simulation methods of controlling pore structures in carbon based nano adsorbents, that focus on parameters such as range of applied temperatures in activation state and applied special catalysts in activation condition that affect controlling of pore structure of carbon materials. This field of research has a great room for improvement in pore structure control modeling and simulation methods of carbon materials. There is no definite answer to this argument since each of these adsorbents has its own advantages and disadvantages in their special applications.

1.2 MATHEMATICAL MODELING FOR PSD AND ADSORPTION PARAMETERS CALCULATING IN CARBON NANO ADSORBENTS

Finding a reliable, accurate, and flexible method for the determination of PSD of porous adsorbents still remains an important concern in the area of characterization of porous materials. Although a large number of researches have been done in this area, some constraints such as type of adsorbate, adsorbent characteristics, adsorption temperature, applicable range of pore size, and range of relative pressure limit the applicability of each model in all cases. The lack of such method is tangible by rapid development of new porous materials and their wide applications in various fields. In the present study, the following three well-known models were used in order to obtain PSD for two series of chemically activated carbons and the results are compared. It is increasingly common to study adsorption processes, whether on free surfaces or in confined spaces such as pores by modeling or simulation techniques. The reach aim of these studies is frequently to develop an understanding that will better enable adsorption measurements to be used to characterize various adsorbents in terms of their surface properties or pore structure. Modeling and simulation methods include: Grand Canonical Monte Carlo (GCMC), density functional theory (DFT), LJ potential, BJH, Novel Olgorithms method such ASA, Verlet, SHN, HK and IHK method, DR method, DS method (Stoeckeli method), etc. [121].

There is also a different approach based on a single adsorption isotherm. Here, total adsorption amount, which is simply a summation of the adsorbed molecules on various adsorption sites, is equal to an integral of the local adsorption on particular sites multiplied by a PSD function, integrated over all sizes:

$$\theta(P) = \int_0^\infty \theta(L, P) f(L) dL \tag{1}$$

where $\theta(L, P)$ is the local adsorption isotherm (kernel) evaluated at bulk pressure P and local pore size (L), and F(L) denotes the PSD of the heterogeneous solid adsorbent. Solving for the PSD function is an ill-posed problem unless the form of function is defined. Various models by assuming different kernels (Langmuir, Freundlich, BET, DR, DA, Sips, Toth, Unilan, Jovanovich, Fowler and Harkins) and mathematical functions (Gaussian, Gamma) for PSD have been presented.

For instance, the Dubinin-Stoeckli (DS) and Stoeckli models which have been proposed based on the Dubinin theory of volume filling of micropores (TVFM) implement Gaussian and gamma type of mathematical function, respectively [1, 2, 133].

1.2.1. BASIC MODELS FOR ADSORPTION PARAMETERS CALCULATING IN CARBON NANO ADSORBENTS

1.2.1.1   CALCULATE OF S_{BET}, V_{MICRO} AND S_{MESO} IN POROUS CARBONS

First stage in characterization of adsorption properties of active carbons is usually determination of their surface area and pore volume. The surface area is normally determined from equilibrium adsorption isotherm of a gas or vapor measured in a range of relative pressures from 0.01 to 0.3. Currently, there are two major methods used to evaluate specific surface area from gas adsorption data: the Brunauer-Emmett-Teller (BET) method and the comparative plot analysis.

The evaluation of the specific area by the BET method is based on the determination of the monolayer capacity (i.e., the number of adsorbed molecules in the monolayer on the surface of a material) by fitting experimental gas adsorption data to the BET equation:

$$a = \frac{a_m C \dfrac{p}{p_0}}{\left(1 - \dfrac{p}{p_0}\right)\left[1 + (C - 1)\dfrac{p}{p_0}\right]} \tag{2}$$

where is the total amount adsorbed, is the monolayer capacity, p/p_0 is the relative pressure and is the constant related to the heat of first-layer adsorption:

$$C = \exp\left(\frac{q_1 - q_L}{RT}\right) \tag{3}$$

where is the difference between the heat of adsorption in the first layer and heat of condensation, T is the absolute temperature and R is the universal gas constant. The above equation (Eq. (3)) was derived for an infinite number of adsorbed layers. This equation is usually expressed in the following linear form:

$$\frac{\dfrac{p}{p_0}}{a\left(1 - \dfrac{p}{p_0}\right)} = \frac{1}{a_m C} + \frac{C - 1}{a_m C}\frac{p}{p_0} \tag{4}$$

Eq. (4) is used to evaluation the monolayer capacity, am, which is necessary for

the evaluation of the surface area, . If the crosssectional area for a single molecule in the monolayerformed on a given surface is known, the surface area can be evaluated by using thefollowing formulae:

$$S_{BET} = a_m \omega N_A$$

(5)

where N_A is the Avogadro number. The derivation of the BET equation involves the following major assumptions: the surface is flat, all adsorption sites exhibit the same adsorption energy; there are no lateral interactions between adsorbed molecules; the adsorption energy for all molecules except those in the first layer is equal to the liquefaction energy; and an infinite number of layers can be formed. In the case of adsorption on active carbons, some of these assumptions are often not valid. In particular, surfaces are geometrically and energetically heterogeneous, there are lateral interactions between adsorbed molecules, and interactions of adsorbed molecules vary with the distance from the surface. Therefore, one should not expect the monolayer capacity evaluated by the BET method to be particularly accurate. In addition, the available values of cross sectional area, even in the case of the most commonly used adsorbates, are somewhat uncertain and may actually vary from one type of the surface to another [1, 2].

Moreover, the determination of the specific surface area based on the molecular size and monolayer capacity should be treated with some caution, because the ability of molecules to effectively cover the surface depends on the molecular size and surface roughness. Since adsorbed molecules cannot satisfactory probe the surface on the scale smaller than their size, the surface area determined using larger molecules might be smaller than that obtained from adsorption data for smaller molecules. Despite all these problems and limitations, the BET method is currently a standard way for evaluation of the specific surface area of solids. For several reasons, nitrogen (at 77 K) is generally considered to be the most suitable adsorbate for surface area evaluation and it is usually assumed that the nitrogen monolayer is close packed. For many years nitrogen adsorption data at 77 K have been used to characterize the porous structures of a variety of materials and reference nitrogen adsorption isotherms for different nanoporous carbons have been reported. However, it is possible to characterize nanoporous carbons by using other adsorbates, some of whichmay be more convenient to use and provide a better insight into the porous structures. Consequently, reference adsorption data on carbons have been published for argon and n-butane, benzene, neo pentane and methanol. In particular, argon deserves much attention, because it was found to be convenient for the characterization of microporous and mesoporous. The intercept on the adsorption axis evaluated by the back extrapolation of the -plot- provides the micropore adsorption capacity :

$$a\left(\frac{p}{p_0}\right) = a_{mi}^0 + \eta\alpha_s$$

(6)

The micropore volume V_{mi} can be evaluated according to the formula:

$$V_{mi} = a_{mi}^0 F$$

(7)

In latter equation, F is a conversion factor (F = 0.0015468 for nitrogen at 77 K, F = 0.001279 for argon at 77 K and 87 K, when the amount adsorbed is expressed in cm ^{3} g^{-1} and the pore volume is expressed in cm^3 g^{-1}). The slope of the linear segment of the -plot () permits the evaluation of the monolayer adsorption capacity for the mesopore surface:

$$a_{me}^0 = \eta\frac{a_s^0}{a_s(0.4)}$$

(8)

where is the monolayer capacity of the reference adsorbent, as(0.4) is the amount adsorbed on the non-porous reference adsorbent at relative pressure p/p$_o$ = 0.4. The mesopore surface area is obtained by multiplying by Avogadro's number N_A and the molecular area occupied by one molecule adsorbed on the mesopore surface, that is:

$$S_{me} = a_{me}^0 N_A \omega$$

(9)

The total surface area, S$_t$, is assessed from the slope of the low-pressure part of the -plot:

$$S_t = \kappa\frac{a_s^0}{a_s(0.4)}N_A\omega$$

(10)

The difference between the total surface area and the mesopore surface area is often used to estimate the surface area of micropores, :

$$S_{mi} = S_t - S_{me}$$

(11)

The -plot method provides an effective and simple way for evaluation of the micropore volume V_{mi}, the total surface area S$_t$ and the mesopore surface area S$_{me}$ of nanoporous materials. The -plots for the active carbons studied obtained by using nitrogen reference adsorption data for non-graphized carbon structures. The slope of the dashed line was used to calculate the total specific surface area, S$_t$, the intercept of the dotted line was used to evaluate the micropore volume, V$_{mi}$, and the slope of the dotted line was used to evaluate the mesopore surface area, S$_{me}$[1, 2, 121, 145].

1.2.1.2 THEORETICAL MODELS FOR ADSORPTION POTENTIAL DISTRIBUTION CALCULATING

The amount adsorbed a (after conversion to the adsorbate volume V) measures the pore volume accessible to adsorption. If V_t is the maximum volume adsorbed, the difference Vt– Vre presents the unoccupied pore volume associated with the adsorption potentials smaller than A, and denotes the non-normalized integral distribution function of the adsorption potential, X*(A). Its first derivative with respect to A is the non-normalized differential distribution function X(A):

$$X(A) = \frac{dX_n^*(A)}{dA} = \frac{d(V_t - V)}{dA} = \frac{dV(A)}{dA}$$

(12)

In terms of the condensation approximation the adsorption potential distribution (APD), X(A), gives essentially the same information as the distribution function of the adsorption energy. Because for microporous carbons the amount adsorbed plotted against the adsorption potential is a temperature independent function, the use of Eq. (12) for calculation of APD is fully justified. The distribution function X(A) can be calculated by numerical differentiation of the characteristic adsorption curve V(A), which is obtained by plotting the amount, adsorbed as a function of the adsorption potential. Previous experimental and theoretical studies of nitrogen adsorption on active carbons indicate some opportunities for using APD for characterization of these adsorbents. It was shown that the minimum or inflection point on the nitrogen APD curve located at about4 kJ mol^{-1} before the peak representing the monolayer formation), can be used to evaluate the monolayer capacity. However, the other minimum, whichappears in the range of lower adsorption potentials, determines the micropore volume. Thus, a typical APD curve for nitrogen on a microporous carbon at 77 Kexhibits two distinct peaks, which represent the monolayer formation (the peak located between 4 and 8 kJ mol^{-1}) and the subsequent volume filling of micropores (the peak between 0.5 and 3 kJ mol^{-1}). The APD curves for nitrogen on the micro and mesoporous carbons at 77K are decreasing functions of A and exhibit an inflection point in the range between 3 and 5kJ/mol, which reflects the completion of the monolayer formation. A sharp increase of the APD curve with decreasing value of A from two to zero reflects the capillary condensation in mesopores. A comparison of the total specific surface area and those evaluated by the BET method and by the -plot method (shows that the APDmethod gives much smaller values. The values of $S_{t,X(A)}$ seem to be more realistic than those obtained by the BET method and -plot analysis, which are based on the BET model, because in contrast to the BET model the APD method allows for a more accurate estimation of the amount adsorbed in the monolayer. The methods based on the BET model do not take into account the correction for molecules adsorbed inside micropores and therefore, they overestimate the total specific surface area of microporous solids, where is the relative adsorption and is the thermal coefficient of the limiting adsorption taken with the minus sign. We could be expressed in terms of the characteristic adsorption curve and the adsorption potential distribution:

$$\ddot{A}S = \left(\frac{\partial A}{\partial T}\right)_\theta - \alpha \frac{\theta(A)}{X(A)} \tag{13}$$

This function provides quantitative information about the distribution of the Gibbs free energy for a heterogeneous porous solid and through Eq. (13) it allows one to estimate the entropy and enthalpy [1, 2, 16, 24, 37].

1.2.1.2.1 LENARD–JONES POTENTIAL FUNCTION

In this work, all of the particles include hydrogen molecules, carbon monoxide molecules and carbon atoms are treated as structure less spheres. Particle–particle interactions between them are modeled with Lenard–Jones potential located at the mass-center of the particles. For a pair of particles i and j separated by the distance r, the interaction between them is given by:

$$\phi_{ij}(r) = 4\varepsilon_{ij}\left[\left(\frac{\sigma_{ij}}{r}\right)^{12} - \left(\frac{\sigma_{ij}}{r}\right)^{6}\right] \tag{14}$$

where i and j donate hydrogen, or carbon monoxide, or carbon particles, ε and σ are the energy and size potential parameters, which are 36.7K and 0.296 nm for hydrogen, 100.2K and 0.3763 nm for carbon monoxide, and 28.2K and 0.335 nm for carbon, respectively. Lorentz–Berthelot rules are used to calculate the parameters of interaction between different kinds of particles. The general potential between a gas molecule and the nanotube is calculated by summing up pair interactions between individual carbon atoms and the gas molecules:

$$V(r) = \sum_j \phi_{ij}(|r_i - r_j|) \tag{15}$$

where r_i is the position of hydrogen or carbon monoxide molecule, r_j is that of a carbon atom and φij (r) is the pair potential between a certain gas molecule and a carbon atom, while the sum is over all of the atoms on the tube. Assuming that the atoms of the solid are distributed continuously up and on a sequence of parallel surfaces that form the pore wall, the interaction potential of the fluid molecule with one of these surfaces of area A and number density θ is given by:

$$V(r) = \int_A \theta v(r)\, d\alpha = 4\varepsilon_{FC}\theta\int_A \left[\left(\frac{\sigma_{FC}}{r}\right)^{12} - \left(\frac{\sigma_{FC}}{r}\right)^{6}\right] d\alpha \tag{16}$$

where $\theta = 38$ nm^{-2}, ε_{FC} and σ_{FC} are Lenard–Jones parameters of the interaction between fluid molecule and carbon atom. The fluids include hydrogen and carbon monoxide. Integrating over the whole nanotube, the following expression can be obtained:

$$V(r,R) = 3\pi\theta\varepsilon_{FC}\sigma_{FC}^2\left[\frac{21}{32}\left(\frac{\sigma_{FC}}{R}\right)^{10} M_{11}(x) - \left(\frac{\sigma_{FC}}{R}\right)^4 M_5(x)\right] \tag{17}$$

where r is the distance between the an atom and the nearest point on the cylinder, R the radius of the nanotube, $x = r/R$ the ratio of distance to radius and θ the same surface number density as above. Here the following integrals are used:

$$M_n(x) = \int_0^\pi \frac{1}{\left(1 + x^2 - 2x\cos\cos\varphi\right)^{n2}} d\varphi \tag{18}$$

Simpson integration is used to get the final potential. For simplicity of programming, the following expression was adopted to represent the results of the integration

$$\frac{V(r,R)}{\varepsilon} = a\left(\frac{\sigma^{10}}{R^{10}}\right) + b\left(\frac{\sigma^4}{R^4}\right) + c\left(\frac{\sigma^4}{R^3}\right) + d\left(\frac{\sigma^4}{R^2}\right) + e\left(\frac{\sigma^4}{R}\right) \tag{19}$$

where a–e are constants dependent on the kind of fluid and the R radius of the nanotubes. They can be determined from the numerical integration. To binary system, the selectivity is defined as

$$S = \frac{x_1/y_2}{x_2/y_2} = \frac{\rho_{p2}^*\rho_{b1}^*}{\rho_{p1}^*\rho_{b2}^*} \tag{20}$$

where x refers to a pore mole fraction and y to a bulk mole fraction. The subscripts of x and y refers to different components in the mixture. is the reduced number density of the bulk fluid. is that of the pore fluid [1, 131, 139, 143, 144, 145].

1.2.1.3 THEORETICAL MODELS FOR ADSORPTION INTEGRAL CALCULATING

Strongly activated carbons possess a broad distribution of micropores because some walls between adjacent micropores burn off. In this case the DR and DA equations cannot give a satisfactory representation of adsorption data. Based on the equations mentioned before that describe adsorption in uniform micropores, Izotova and Dubinin proposed the following two-term equation for description of adsorption on solids with bimodal microporous structure:

$$a_{mi} = a_{mi}^{0I}\exp\left[-B_I\left(\frac{A}{\beta}\right)^2\right] + a_{mi}^{0II}\exp\left[-B_{II}\left(\frac{A}{\beta}\right)^2\right] \tag{21}$$

Here parameters and correspond to micropores and and correspond to super micropores. Eq. (21) has been often applied to describe adsorption on non- uniform microporous solids. The model studies showed that Eq. (21) is especially suitable for description of adsorption on microporous solids that possess two types of micropores of considerably different sizes. For microporous solids with a great number of micropores of different sizes, the summation in Eq. (21) should be replaced by integration and then a_{mi} is given by:

$$a_{mi} = a_{mi}^0 \int_0^\infty \exp\left[-B\left(\frac{A}{\beta}\right)^2\right] F(B) dB \tag{22}$$

where F(B) is the distribution function of the structural parameter B normalized to unity. The integral Eq. (22) was first proposed by Stoeckli. Expressing in this integral the structural parameter B by means of the half-width x:

$$B = \varsigma x^2 \tag{23}$$

Where is the proportionality constant; we have:

$$a_{mi} = a_{mi}^0 \int_0^\infty \exp\left(-mx^2 A^2\right) J(x) dx \tag{24}$$

where is the micropore size distribution and m is defined by:

$$m = \frac{\varsigma}{\beta^2} \tag{25}$$

Comparison of the integral Eqs. (22 and (24) gives the following relationship between the distribution functions F(B) and J(x):

$$J(x) = 2\varsigma x F\left[B(x)\right] \tag{26}$$

$$X_{mi}(A) = 2A\beta^{-2} m \int_0^\infty x^2 \exp\left[-mx^2 A^2\right] J(x) dx \tag{27}$$

Eqs. (26) and (27) define the relationship between distribution function $X_{mi}(A)$, F(B) and J(x).The average adsorption potential associated with Eqs. (26) and (27) is given by:

$$\bar{A} = \frac{\beta\sqrt{\pi}}{2} \int_0^\infty \frac{F(B)}{\sqrt{B}} dB = \frac{1}{2}\left(\frac{\pi}{m}\right)^{1/2} \int_0^\infty \frac{J(x)}{x} dx \tag{28}$$

The dispersion for the distribution function $X_{mi}(A)$ may be expressed as follows:

$$\sigma_A = \left[\beta^2 \int_0^\infty \frac{F(B)}{B} dB - \bar{A}^2\right]^{1/2} = \left[\frac{1}{m} \int_0^\infty \frac{J(x)}{x} dx - \bar{A}^2\right]^{1/2} \tag{29}$$

where A is defined by Eqs. (28) and (29) have general character and permit calculation of A and for arbitrary micropore distributions F(B) and J(x)[1, 2].

1.2.2 BASIC MODELS FOR PSD CALCULATING FROM DATA OF SIMULATED ADSORPTION ISOTHERMS

1.2.2.1 THEORETICAL MODELS FOR MAIN TYPES OF COMPARATIVE PLOTS THAT CALCULATING PSD

Physical adsorption of gases and vapors on a non-porous surface or on the mesopore surface occurs via layer-by-layer mechanism, whereas adsorption in micropores resembles the volume filling mechanism. In the case of porous solids containing both micropores and mesopores, active carbons, ordered microporous and mesoporous carbons and active carbon fibers, the volume filling of micropores occurs first at low pressures and it is followed by the formation of a multilayer film on the mesopore walls, and finally, the remaining empty space inside mesopores is filled via capillary condensation. Thus, the dependence of the amount adsorbed on a porous solid plotted (compared) against the amount adsorbed on reference non-porous solids is linear at higher pressures because the layer-by-layer adsorption occurs on both solid surfaces. However, at low pressures the adsorption mechanisms on the solid studied and on the reference adsorbent can be the same or different, resulting in linear or non-linear behavior of the initial segment of the comparative plot. It should be noted that the linear segment of the comparative plot at higher pressures is rather insensitive on the choice of the reference solid because after all micropores are filled and the first adsorbed layer is completed, the surface effects are negligible and the film formation is mostly controlled by adsorbate-adsorbate interactions. The slope of the linear segment at the low pressures is proportional to the total surface area and the slope of the linear segment at high pressures is proportional to the external surface area of mesopores, whereas its intercept determines the maximum amount adsorbed in micropores, which can be converted to the micropore volume. There are several types of comparative plots such as the t-plot, -plot and -plot, which differ only in the way of presenting the standard adsorption isotherm measured on the reference solid. In the case of the-plot, the standard isotherm isexpressed in terms of the surface coverage, which is the ratio of the amount adsorbed to the monolayer capacity. The thickness of the surface film on the reference solid,, whichis obtained by multiplication of the surface coverage by the monolayer thickness, is used toconstruct the t-plot. In the-method the amount adsorbed on a porous solid is plotted against the reduced standard adsorption. The -method was proposed. A brief description of this method is provided below. The total amount adsorbed at relative pressure is the sum of the amount adsorbed in the micropores and the amount adsorbed on the mesopore surface:

$$a\left(\frac{p}{p_0}\right) = a_{mi}^0 \theta_{mi}\left(\frac{p}{p_0}\right) + a_{me}^0 \theta\left(\frac{p}{p_0}\right) \qquad (30)$$

Here, denotes the maximum amount adsorbed in the micropores, and denotes themonolayer capacity of the mesopore surface. For the standard adsorption isotherm measuredon a non-porous reference adsorbent, we have:

$$a_s = \frac{a_s \left(\frac{p}{p_0} \right)}{a_s (0.4)}$$

(31)

Here, is the relative surface coverage for a non-porous reference adsorbent, isthe amount adsorbed on the surface of this reference adsorbent at relative pressure denotes the monolayer capacity evaluated from the standard adsorption isotherm, and denotes the amount adsorbed on the surface of the reference solid at relative pressure and we have:

$$a \left(\frac{p}{p_0} \right) = a_{mi}^0 \theta_{mi} \left(\frac{p}{p_0} \right) + \eta \alpha_s$$

(32)

$$\eta = \frac{a_{me}^0 a_s (0.4)}{a_s^0}$$

(33)

Kelvin-type equation is usually used to relate the capillary condensation pressure to the radius of cylindrical pores:

$$r \left(\frac{p}{p_0} \right) = \frac{2 \gamma V_m}{RT \ln \ln \left(\frac{p_0}{p} \right)} + t \left(\frac{p}{p_0} \right)$$

(34)

This comparison involved t-curves proposed by Harkins and Jura:

$$t \left(\frac{p}{p_0} \right) = 0.1 \left[\frac{13.99}{0.034 - \log \left(\frac{p_0}{p} \right)} \right]^{0.5}$$

(35)

and Halsey model pressure

$$\left(\frac{p}{p_0} \right) = 0.354 \left[\frac{-5}{\ln \ln \left(\frac{p_0}{p} \right)} \right]^{0.333}$$

(36)

where $t(p/p_0)$ is the statistical film thickness (in nm) for nitrogen adsorbed on the carbon surface and p/p_0 is the relative pressure. It was shown that the applicability of the Barrett, Joyner and Halenda (BJH) computational method based on the

Kelvin equation could be extended significantly towards small mesopores and large micropores when a proper t-curve was used to represent the film thickness of nitrogen adsorbed on the carbon surface. The t-curve proposed in the work gave the pore-size distribution functions for the carbons studied that reproduce the total pore volume and show realistic behavior in the range at the borderline between micropores and mesopores [1, 2, 16, 24, 37, 121, 145].

1.2.2.2 DUBININ-STOECKLI (DS) EQUATION

The adsorption of vapors by microporous carbons was described by the following fundamental equation of Dubinin-Astakhov (DA):

$$W = W_0 \exp\exp\left[-\left(\frac{A}{E}\right)^n\right] \tag{37}$$

Here, W (mmol·g^{-1}) represents the amount adsorbed at relative pressure P_0/P, W_0 denotes the limiting amount of micropores filling, and A is the differential molar work of adsorption defined as A = RT ln(P_0/P) at temperature of T. One may write that E = βE_0, where β is the affinity coefficient depending on the adsorptive only, and it has been assumed that for benzene as a reference β = 1. In general case of heterogeneous microporous adsorbents, the adsorption is described by the Dubinin-Stoeckli (DS) adsorption equation of:

$$W = \frac{W_0}{2\sqrt{1+2m\delta^2 A^2}}\exp\left(-\frac{mx_0^2 A^2}{1+2m\delta^2 A^2}\right)\times\left[1+erf\left(\frac{x_0}{\delta\sqrt{2}\sqrt{1+2m\delta^2 A^2}}\right)\right] \tag{38}$$

which implies a normal half-width (x) distribution of micropore volume for the slit-like pores as:

$$\frac{dW}{dx} = \frac{W_0}{\delta\sqrt{2\pi}}\exp\left[-\frac{\left(x-x_0\right)^2}{2\delta^2}\right] \tag{39}$$

where, x_0 is the half-width of a slit shaped micropore, which corresponds to the maximum of the distribution curve, and δ is the variance. The letter m is a constant coefficient for a given vapor:

$$m = \left(\frac{1}{\beta k}\right)^2 \tag{40}$$

For benzene as the reference vapor, the constant k equals to 12 kJ·nm·mole^{-1}. Using Eq. (38) to fit the experimental data, three parameters of W_0, x_0 and δ, can be extracted. Knowing these parameters, the micropore size distribution in terms of volume can be calculated from Eq. (39) [1, 2, 19, 126, 133].

1.2.2.3   STOECKLI MODEL

Another approach for the determination of PSD of porous adsorbents which is also based on the Dubinin's TVFM is Stoeckli method. It had been shown by scientists, that for the ideal slit-shaped microporous materials, a good estimate of the adsorption isotherm can be obtained by:

$$W = W_0 \left[\frac{a}{a + \left(A / \beta K_0 \right)^3} \right] \tag{41}$$

$$\frac{dW}{dL} = \frac{3W_0 a^v L^{3v-1} \exp\exp\left(-aL^3\right)}{\tilde{A}(v)} \tag{42}$$

The a and v are constant parameters that are related to the mean and width of the distribution, respectively. K_0 is calculated using Eq. (42). This is applicable over a range of pore size from0.4 to 2.0 nm. After obtaining the model parameters using Eq. (41), pore size distribution can be determined using the following gamma type distribution of the mean pore width (L = 2x) [1, 126, 133]:

$$\frac{dW}{dL} = \frac{3W_0 a^v L^{3v-1} \exp\exp\left(-aL^3\right)}{\tilde{A}(v)} \tag{43}$$

1.2.2.4   HORVATH-KAWAZOE (HK) METHOD

Horvath and Kawazoe developed a rather simple means of characterizing the pore structure of porous materials. This model provides a simple, one-to-one correspondence between the pore size and relative pressure at which the pore is filled. Using thermodynamic arguments and applying the potential obtained by Horvath and Kawazoe derived the following expression:

$$RT \ln \ln \left(\frac{P}{P_0} \right) + \left[RT - \frac{RT}{\theta} \ln \ln \frac{1}{1-\theta} \right] = N_{AV} \frac{N_a A_a + N_A A_A}{\delta^4 (L - 2d_0)} \times \left[\frac{\delta^4}{3(L-d_0)^3} - \frac{\delta^{10}}{9(L-d_0)^9} - \frac{\delta^4}{3d_0^3} + \frac{\delta^{10}}{9d_0^9} \right]$$

(44)

where L represents the micropore width (L = 2x), N_{AV} denotes Avogadro's number, and R and T are gas constant and temperature, respectively. A_a and A_A are dispersion constant characterizing adsorbate-adsorbent and adsorbate adsorbate interactions, N_a and N_A are the number of atoms per unit area of adsorbent and the number of molecules per unit area of adsorbate, d_0 is the arithmetic mean of the adsorbate molecular diameter and the adsorbent atomic diameter, and δ is the distance between a gas molecule and an adsorbent atom at zero interaction energy at relative pressure of P/P_0. From the amount adsorbed at relative pressure of P/P_0, Eq. (44) yields the corresponding slit-pore width, L. Thus, a plot of adsorbed volume versus L is a cumulative pore-volume curve, the slopes of which give the differential PSD [133].

1.2.2.5   IMPROVED HORVATH-KAWAZOE (IHK) METHOD

The HK equation is widely used for calculating the micropore size distribution (MPSD) from a single adsorption isotherm measured at subcritical temperature (e.g., N_2 at 77 K). In the HK model, the ideal Henry's law (or linear behavior) is assumed for the isotherm. Cheng and Yang [51] modified the HK formulation by assuming the non-linearity of the isotherm equation. This has improved the HK model significantly with the advantage of maintaining the PSD calculation simple. The non-linearity assumption also results in sharpening PSD. Considering the mentioned non-linearity correction, Cheng and Yang [51] derived the improved HK equation (IHK) for three different pore geometries. For the slit-shaped pores, the IHK equation is derived as:

$$\theta = \frac{C_\mu}{C_{\mu s}} = \exp\left[-\frac{1}{\left(\beta E_0\right)^2}\left(R_g T \ln \ln \frac{P}{P_0} \right)^2 \right] \tag{45}$$

The influence of this term and thus θ depends on P/P_0 where the adsorption occurs and also the shape of the isotherm. In the initial part of the adsorption isotherm, where θ is small and θ-dependent term (second term in LHS of IHK equation) approaches zero, the IHK model approaches HK. By increasing θ, the pore filling term $RT - (RT/\theta) \ln[1/(1 - \theta)]$ becomes more negative. In the meantime, as the relative pressure is increased, the free energy term $RT \ln(P/P_0)$ increases. The increase in free energy term is partly offset by the pore filling term. So, the LHS of the IHK equation is increased at a slower rate as compared with the original HK equation. Consequently, the calculated pore size is increased at a slower rate, resulting in sharpening the pore size distribution. In the present study, θ has been calculated from the f DR equation [133].

$$\theta = \frac{C_\mu}{C_{\mu s}} = \exp\left[-\frac{1}{\left(\beta E_0\right)^2}\left(R_g T \ln \ln \frac{P}{P_0} \right)^2 \right] \tag{46}$$

1.2.2.6   MOLECULAR SIMULATION POROSIMETRY

The experimental adsorption isotherm measured on a porous solid sample is the aggregate of the isotherms for the individual pores of different sizes. Consequently, the experimental isotherm is the integral of the single pore isotherm multiplied by the pore size distribution, if we neglect the geometrical and chemical heterogeneities in the porous surfaces. For a slit-shaped pore, this can be described as:

$$N(P) = \int_{H_{min}}^{H_{max}} f(H)\rho(P,H)\,dH \tag{47}$$

where N(P) is the amount adsorbed at pressure p, H_{min} and H_{max} are the widths of the smallest and largest pores, (P,H) is the mean density of N_2 at pressure P in a pore of width H. The N(P) versus P relation is just an adsorption isotherm. f(H) is a pore size distribution function, the distribution of pore volumes as a function of pore width H. Therefore, all of the heterogeneities of less crystalline porous solids are approximated by the distribution of pore sizes. If (P,H) can be obtained from the molecular statistics, f(H) can be determined by the best lit to the observed experimental isotherm. The width H in f(H) is not the effective pore width that as mentioned in above models. In order to derive the molecular density in a pore, statistical approaches to fluids have been used. Seaton et al. applied the mean field theory to calculate (P,H). The mean field theory is an approximate theory of inhomogeneous fluids in which the interactions between the fluid molecules are divided into a short-range, repulsive part and a long-range, attractive part. Each is treated separately for faster calculation than full molecular simulation. The contribution of the long-range forces to the fluid properties is treated in the mean field approximation, while the effect of the short-range forces is modeled by an equivalent array of hard spheres. There are two approaches to get the short-range forces–the Local mean field theory and the non-local one, where the former neglects the short-range correlation, bat the latter takes it into account. Seaton et al. adopted the local density approach for their calculation. They calculated (P,H) by the above method. How can we determine f(H) from the calculated (P, H) and the experimental adsorption isotherm N(P)? It has a mathematical difficulty. They used the following bimodal log-normaldistribution, which is flexible to represent the various pore size distributions and is zero for all negative pore widths:

$$f(H)=\left\{V_1\left/\left[\sigma H(2\pi)^{1/2}\right]\right\}\exp\left\{-\left[\log H-\mu_1\right]^2/2\sigma_1^2\right\}+\left\{V_2\left/\left[\sigma_2 H(2\pi)^{1/2}\right]\right\}\times\exp\left\{-\left[\log H-\mu_2\right]^2/2\sigma_2^2\right\}\ (48)$$

where Vi is the pore volume of the distribution i, and a, and, u, are the parameters defining the distribution shape. These six parameters in Eq. (48) are determined from the best fit to the experimental adsorption isotherm. The limit of H_{mi} corresponds to the smallest pore into which the N_2 molecule can enter. On the other hand, the upper limit of H, is determined by the width of the mesopore, which condenses at the highest experimental pressure. This calculation can determine the pore size distribution from micropore to mesopore. In that work the applicability for the pores of less than 1.3 nm was not shown. Lastoskie et al. extended the above method to the non-local mean field theory. The non-local mean field theory gives a quantitative accurate description of even ultra micropore structures. They compared the pore size distributions from the local and non-local mean field theories as to real adsorption isotherms by activated carbons; the local theory underestimates the pore size distribution compared with the non-local theory. As the calculation with the mean field density theory often gives a qualitative agreement rather than quantitative one. The grand canonical ensemble Monte Carlo simulation is also necessary for such an approach. They also got good results. The molecular simulation studies on the pore size distribution have shown a new picture on the adsorption in the wide range of pores from ultramicropores to mesopores. Understanding of micropore filling and capillary condensation proceeds rapidly.

In the future the pore connectivity will be taken into account, so that a more elaborated method will be settled, although so far molecular simulation porosimetry is not a popular method [1, 2].

1.2.2.7 ADVANCE METHODS BASED ON DENSITY FUNCTIONAL THEORY

Beside classical methods of pore size analysis, there are many advanced methods. Some researchers proposed a method based on the mean field theory. Initially this method was less accurate in the range of small pore sizes, but even so it gave a more realistic way for evaluation of the pore size distribution than the classical methods based on the Kelvin equation. More rigorous methods based on molecular approaches such as grand canonical Monte Carlo (GCMC) simulations and non-local density functional theory (NLDFT) have been developed and their use for pore size analysis of active carbons is continuously growing. Let us consider a one-component fluid confined in a pore of given size and shape, which is itself located within a well-defined solid structure. We suppose that the pore is open and the confined fluid is in thermodynamic equilibrium with the same fluid (gas or liquid) in the bulk state at a given temperature. As the bulk fluid is homogeneous, its chemical potential is simply determined by the pressure and temperature. The fluid in the pore is not of constant density and it is subjected to adsorption forces in the vicinity of the pore walls. This in homogeneous fluid, which is stable under the influence of the external field, is in effect a layer-wise distribution of the adsorbate. The density distribution can be characterized in terms of a density profile, $p(r)$, expressed as a function of distance, r, from the wall across the pore. In the density functional theory (DFT) the statistical mechanical grand canonical ensemble is used. The appropriate free energy quantity is the grand Helmholtz free energy, or grand potential functional, D(r). This free energy functional is expressed in terms of the density profile $p(r)$: then by minimizing the free energy (at constant, V, T) it is possible in principle to obtain the equilibrium density profile. For a one-component fluid, which is under the influence of a spatially varying external potential, the grand potential functional becomes:

$$\Omega\big[\rho(r)\big] = F\big[\rho(r)\big] + \int \rho(r)\big[\Phi(r) - \mu\big] dr \tag{49}$$

where $F[p(r)]$ is the intrinsic Helmholtz free energy functional, is the external potential,and the integration is performed over the pore volume V. The $F[p(r)]$ functional can be separated into an ideal gas term and contributions from the repulsive and attractive forces between the adsorbed molecules the fluid- fluid interactions)[1, 2, 11, 36].

Hard-sphere repulsion and pair wise Lenard-Jones potential are usually assumed and a mean field treatment is generally applied to the long-range attraction. In the earliest local version of density functional theory (LDFT) the Helmholtz free energy was assumed to be a single valued function of the local density $p(r)$.

Further investigations based on comparison of the density profile provided by the LDFT and that determined with GCMC simulations showed that in the case of inhomogeneous fluid more rigorous analysis requires accounting for the density distribution in the region of a few collision diameters in the proximity of a given point. For this reason, it is now customary to apply non-local density functional theory (NLDFT), which involves the incorporation of short-range smoothing functions. In this manner, it has been possible to obtain good agreement with the density profiles determined by Monte Carlo molecular simulations. The non-local density functional theory (NLDFT) is well established and widely presented in the literature. The distribution of density in a confined pore can be obtained for an open system in which a pore is allowed to exchange mass with the surroundings. From the thermodynamic principle, the density distribution is obtained by minimization of the following grand potential written below for the one-dimensional case:

$$\Omega\left[\rho(z)\right] = \int \rho(z)\left[f(z) + \Phi(r) - \mu\right] dz \tag{50}$$

Here $\rho(z)$ is the local density of the adsorbed fluid at a distance z from one of the walls of the pore, $f(z)$ is the intrinsic molecular Helmholtz free energy of the adsorbate phase, is the chemical potential. The flee energy $f(z)$ comprises the ideal, mean-field attractive terms, and the excess free energy (repulsive) term as a function of smoothed weighted average. A new approach based on NLDFT to determine pore size distribution (PSD) of active carbons and energetic heterogeneity of the pore wall was proposed by others. The energetic heterogeneity is modeled with an energy distribution function (EDF), describing the distribution of solid- fluid potential well depth (this distribution is a Dirac delta function for an energetically homogeneous surface). The approach allows simultaneous determining of PSD (assuming slit shape) and EDF (from nitrogen isotherms by using a set of local isotherms calculated for a range of pore widths and solid fluid potential well depths. It was found that the structure of the pore wall surface differs significantly from that of graphitized carbon black. This could be attributed to defects in the crystalline structure of the surface, active oxide centers, finite size of the pore walls (in either wall thickness or pore length), and so forth. Those factors depend on the precursors and the process of carbonization and activation and hence provide a fingerprint for each adsorbent. Ustinov and Do approach gives an accurate representation of the experimental adsorption isotherm. The pore size distributions indicate quite significant differences in the porosity of the carbons studied in the range of micropores and mesopores [1, 11, 36].

1.2.2.8 JARONIEC-CHOMA METHOD
The integral equation that mention before was solved for various continuous functions representing the distribution F(B). For example, Wojsz and Rozwadowski solved the integral equation for distribution functions F(B) other than the Gaussian

one, decreasing and increasing exponential distributions and Rayleigh distribution. Some of these equations have rather complex mathematical form; however, the decreasing exponential and Rayleigh distributions may be considered as special cases of the gamma-type distribution, which generate a very simple isotherm equation. General integral equation similar to integral equation for the adsorption isotherm on heterogeneous microporous solids can be written as follows:

$$a_{mi} = a_{mi}^0 \int_0^\infty \theta_{mi}\left(z, A\right) F(z) dz \qquad (51)$$

where a_{mi} is the equilibrium amount adsorbed in micropores, is the maximum amount adsorbed in micropores, is the local isotherm describing adsorption in uniform micropores; z is the quantity associated with the micropore size; $A = RT$ In (p/p_0) is the adsorption potential, $F(z)$ is the distribution function characterizing heterogeneity of microporous structure. Jaroniec and co-workers proposed the following gamma-type distribution function:

$$F(z) = \left[\frac{n\rho^v}{\Gamma\left(v/n\right)}\right] z^{v-1} \exp\left[-(\rho z)^n\right] \qquad (52)$$

whereis the inverse value of the characteristic energy E_0 for the reference adsorbate, $p > 0$ and $v > 0$ are parameters for the gamma distribution function. It was shown elsewhere that the Jaroniec-Choma (JC) equation, which was obtained by generalization of the DA equation for $n = 2$ or $n = 3$, gives good description of gas and vapor adsorption for many microporous active carbons. A general form of the JC equation can be written as:

$$a_{mi} = a_{mi}^0 \left[1 + \left(\frac{A}{\beta\rho}\right)^n\right]^{-v/n} \qquad (53)$$

Here a_{mi} and denote respectively the amount and the maximum amount adsorbed in micropores, p and v are parameters of the gamma distribution function. Eq. (53) with $n = 2$ was proposed by Jaroniec and Choma on the basis of the assumption that mention before with $n = 2$ [Dubinin-Radushkevich (DR) equation] governs adsorption in uniform micropores. This assumption was justified experimentally by Stoeckli and Dubinin and others. Later, Stoeckli carried out careful adsorption and calorimetric experiments for benzene on molecular carbon sieves, and showed that DA equation with $n = 3$ gives a better representation of adsorption in uniform micropores than that with $n = 2$. These experimental studies suggest that the DA equation with $n = 2$ describes adsorption in nearly uniform micropores and its generalized from [Eq. (53) with $n = 2$] gives a good description of adsorption on microporous active carbons with large structural heterogeneity. For microporous solids with moderate structural heterogeneity, the use of the isotherm equations obtained by generalization of DA equation with n = 3 [the JC Eq. (53) with n = 3] is substantiated better than the use of those generated by DA equation with n = 2. It is noteworthy that some authors postulated that DA equation can be used to describe adsorption in uniform micropores several years before this postulate found some experimental justification. The distribution function F(z) together with the quantities and are used to characterize the structural heterogeneity of microporous solids. Energetic heterogeneity of a microporous solid generated by the overlapping of adsorption forces from the opposite micropore walls can be described by the adsorption potential distribution in micropores . This distribution associated with Eq. (53) is given by:

$$X_{mi}(A) = v(\beta\rho)^{-n} A^{n-1} \left[1 + \left(\frac{A}{\beta\rho} \right)^n \right]^{-\frac{v}{n}-1} \tag{54}$$

The adsorption potential distribution in micropores $X_{mi}(A)$ given by Eq. (54) can be characterized by the following quantities. Although description of microporous structures of nanoporous carbons is a difficult and still not fully solved task, comparative studies of various adsorption models can facilitate elaboration of methods for characterization of microporous solids. It was shown that the gamma distribution function F(z) gives a good description of structural heterogeneity for many microporous carbonaceous materials. For microporous active carbons with small structural heterogeneity the JC equation gives a good description of adsorption in micropores, while the JC equation can be used for adsorption on microporous active carbons with strong structural heterogeneity [1, 2, 19].

1.2.2.9   DUBININ-RADUSHKEVICH (DR) EQUATION

The Dubinin-Radushkevich (DR) equation, proposed in 1947, undoubtedly occupies a central position in the theory of physical adsorption of gases and vapors on microporous solids. The amount adsorbed in micropores a_{mi} is:

$$a_{mi} = a - a_{me} \tag{55}$$

where a is the sum of the amount adsorbed in micropores a_{mi} and in mesopores a_{me}. According to the DR equation a_{mi} is a simple exponential function of the square of the adsorption potential A:

$$a_{mi} = a_{mi}^0 \exp\left[-B \left(\frac{A}{\beta} \right)^2 \right] \tag{56}$$

Here B is the temperature-independent structural parameter associated with the micropore sizes, andbis the similarity coefficient, which reflects the adsorbate properties that such equations is commonly used for description of gas and vapor adsorption on microporous active carbons.

A more general expression is that proposed by Dubinin and Astakhov, which is known as the DA equation: a_{mi}:

$$a_{mi} = a_{mi}^0 \exp\left[-\left(\frac{A}{\beta E_0}\right)^n\right]$$ (57)

In Eqs. (55), (56) and (57) a_{mi} represents the amount adsorbed in micropores at relative pressure p/p_0 and temperature T, is the limiting volume of adsorption or the volume of micropores and Vm is the molar volume of the adsorbate. The specific parameters of Eqs. (55) and (56) are and n, respectively, where E_0 is the characteristic energy of adsorption for the reference vapor, usually benzene. Dubinin proposed to extract the amount ami adsorbed in micropores from the total adsorbed amount $a(p/p_0)$ as follows:

$$a_{mi} = a - S_{me}\gamma_s$$ (58)

The amount adsorbed per unit surface area was evaluated from the adsorption isotherm measured on a reference adsorbent, whereas the specific surface area Sine of mesopores was estimated from the adsorption isotherm. To analyze nitrogen adsorption isotherms by means of the DR and DA equations, we extracted according to Eq. (58) the adsorption isotherm $a_{mi}(p/p_0)$ for micropores from the total adsorption isotherm $a(p/p_0)$. To calculate a_{mi} according to Eq. (58), we evaluatedfrom the standard nitrogen isotherm at 77°Kshown by the -plot method. The extracted adsorption isotherm was described by DR equation (the Eq. (55)) and also by DAequation (the Eq. 56)[1, 19, 126].

1.2.2.10 DFT-NLDFT MODELS BASE ON KERNELS OF EXPERIMENTAL ADSORPTION ISOTHERM DATA

The PSD is calculated from the experimental adsorption isotherm with (P/P_0) by solving the integral adsorption equation:

$$N_{exp}\left(P/P_0\right) = \int_{D_{min}}^{D_{max}} N_{QSDFT}\left(P/P_0, D\right) f(D) dD$$ (59)

The experimental isotherm is represented as the convolution of the DFT kernel (set of the theoretical isotherms NQSDFT(P/P_0,D)in a series of pores within a given range of pore sizes D) and the unknown PSD function f (D), where D_{min} and D_{max} are the minimum and maximum pore sizes in the kernel, respectively. Two kernels of the selected DFT adsorption isotherms for the slit geometry are reviewed, include NLDFT and QSDFT. In contrast to the NLDFT kernels, the QSDFT isotherms are smooth prior to the capillary condensation step, which is characteristic of mesopores (D > 2 nm), and thus do not exhibit stepwise inflections caused by artificial layering transitions Solution of above equation can be obtained using the quick non-negative least square method. It should be noted that in NLDFT kernels, the pore width is defined as the center-to-center distance between the outer layers of adsorption centers on the opposite pore walls corrected for the solid–fluid LJ diameter. In QSDFT kernels, the pore width is defined from the condition of zero solid density excess. These definitions are apparently different, albeit insignificantly, but this difference should be taken into account in data analysis. Over the years a library of NLDFT and more recently QSDFT kernels were developed for calculating pore size distributions in carbonaceous and silica micro mesoporous materials of different origin from nitrogen and argon adsorption isotherms, as well as for microporous carbons from carbon dioxide adsorption. For a DFT kernel of a given adsorbate–adsorbent pair, the parameters should not only represent the specifics of adsorbent–adsorbate interactions, but also take into account the pore structure morphology. DFT kernels were built for calculating PSD using different adsorbates: nitrogen at 77.4 K, argon for 77.4 and 87.3 K, and carbon dioxide at 273 K. It was shown, that the results obtained with different adsorbates are in reasonable agreement. Nitrogen at77.4 K is the conventional adsorbate for adsorption characterization [1, 15, 39].

However, argon and carbon dioxide are more suitable in some cases, especially for microporous materials. In general, argon at 87.3 K is always a better molecular probe than nitrogen, since it does not give rise to specific interactions with a variety of surface functional groups, which can lead to enhanced adsorption/specific interactions caused by quadru pole moment characteristic to non-symmetric molecules. In addition, argon at 87.3 K fills micropores of dimensions 0.5–1 nm at higher relative pressures compared to nitrogen at 77.4 K, and, due to faster diffusion, the equilibrium times are shorter. As such, it is possible to test micropores as small as 0.5 nm with argon within the reliable range of relative pressures that is limited in modern automated instruments. The advantages of using argon are very pronounced for zeolites and metal-organic frameworks Historically, the first DFT kernels were developed for carbon slit pores some researchers designed the consistent equilibrium NLDFT kernels for nitrogen, argon, carbon dioxide isotherms, which are applicable for disordered micro-mesoporous carbons of various origin, including activated carbons, and carbon fibers, charcoal, and carbon black. Activated carbon fibers (ACF) exhibit a type I isotherm and possess a very high adsorption capacity with BET surface areas up to 3000 m^2/g. This results in rapid adsorption and desorption rates with over 90% of the total surface area belonging to micropores of 2 nm or less. And like their granular counterparts, AC Fare finding a foot hole in a broad range of applications including gas and liquid phase adsorption, carbon molecular sieves, catalysis, gas storage, and super-capacitors. The factors that greatly affect the ACF properties (precursor source, temperature, time, gas flow activating agents and the use of catalysts) are the ones that most influence the pore structure. ACF prepared by a physical activation process will be dependent on a controlled gasification process at temperatures ranging from 800o 1000°C. In their activation procedure they applied the NLDFT method and showed that a greater degree of activation led to a widening of the pore size distribution from 2.8 to 7.0 nm. They contribute this broadening to a decrease in the number of micro domains. This phenomenon was coupled with an increase in the peak pore size (from 0.44 nm to 1.86). The adsorption data and sub sequential pore size analysis was confirmed by NMR. The chemical activation process on the other hand involves the mixing of a carbon precursor with a chemical activating agent typically KOH, NaOH, H_3PO_4 or $ZnCl_2$ [1, 127, 129, 142].

1.2.2.11 ADSORPTION INDUCED MOLECULAR TRAPPING (AIMT) IN MICROPOROUS MEMBRANE MODEL

Model and simulation schemes and Principles of the DCV-GCMD numerical experiment in microporous membrane is illustrated in Fig. 1.5.

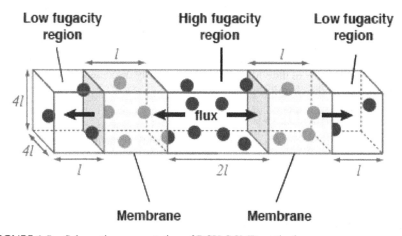

FIGURE 1.5   Schematic representation of DCV-GCMD method.

In our simulations the membrane thickness stands as one unit cell: l = 2:5 nm. Our numerical experiment consists in reproducing an experimental set-up used for permeability measurements, as illustrated in below detailed. For that purpose we use the DCV-GCMD method with high and low fugacity reservoirs are imposed at each end of the membranes. This allows the application of periodic boundary conditions. While the fugacities in these reservoirs are controlled by means of Grand Canonical Monte Carlo simulation, molecular motions are described using Molecular Dynamics simulation. Once the system has reached the steady-state regime, the molar flux J is estimated by counting the number of molecules N crossing the membrane of cross section S during a time interval :

$$J = \frac{N}{S\Delta t} \qquad (60)$$

The permeability is then defined as the flux per unit of fugacity gradient ($D_f = 1$) across the membrane:

$$P = \frac{Jl}{\Delta f} \qquad (61)$$

From the thermo dynamical point of view, a system of fluid particles sorbed in a immobile porous medium deviates from equilibrium when a gradient in molecules chemical potential exists. Under isothermal conditions and on the assumption that transport mechanism is diffusive, the local molar flux ~j (number of moles of fluid per unit surface per unit time) through the fixed porous solid satisfies the Maxwell-Stefan equation:

$$\vec{j} = -\frac{cD_o}{RT}\vec{\nabla}\mu \qquad (62)$$

where T is the temperature, R the ideal gas constant, c the average interstitial concentration(number of moles per material unit volume), while Do stands as the collective diffusivity of the sorbed fluid, as previously discussed. Moreover, using the definition of the chemical potential:

$$\mu(f,T) \equiv \mu_0(T) + RT \ln\ln(f/f_0) \qquad (63)$$

one can alternatively consider the gradient in fugacity as the driving force of fluid motion. Hence the rearranged expression of the local molar flux:

$$\vec{j} = -\frac{cD_o}{f}\vec{\nabla}f \qquad (64)$$

where f is the fluid fugacity. In order to estimate the concentration in the microporous membrane, refer to the classical Langmuir model, commonly used to describe adsorption isotherms of fluids in microporous adsorbents:

$$c = c_s \frac{bf}{1+bf} \qquad (65)$$

Here is the complete filling concentration and is an equilibrium adsorption constant, whichcan be interpreted as the inverse of a characteristic filling pressure. We stress that we use the Langmuir model for its ability to reproduce the adsorption isotherms simulated in our membrane models (see supplementary information) and its convenient analytical form. Let us now consider a microporous membrane of thickness in the direction and separating two infinite bulk fluid reservoirs exhibiting a difference in chemical potentials. Under these conditions Eq.

(65), describes the local motion of interstitial fluid in the membrane. In an actual experiment, one can only measure the total molar flux as a function of the fugacity drop across the membrane. Hence the averaged transport equation, obtained from the average of this equation along the thickness of the membrane:

$$\vec{j} = -P_e \frac{\Delta f}{l} \vec{e}_x = -\frac{1}{l} \int_{x=0}^{x=l} dx \frac{D_o c}{f} \left(\frac{\partial f}{\partial x} \right) \vec{e}_x \qquad (66)$$

in which is referred to as the permeability of the membrane. It should be stressed that this definition of the permeability differs from the classical definition deriving from Darcy's law, which considers the viscous flow of a Newtonian interstitial fluid. In the present case, for the sake of generality we define the permeability as the transport coefficient relating the molar flux to the driving force of fugacity gradient, as found in the literature, Assuming a constant , we deduce the overall permeability of the membrane from Eq.(67) as:

$$P_e = \frac{D_o}{\Delta f} \int_{f_u}^{f_d} df \frac{c}{f} \qquad (67)$$

in which and are the downstream and upstream fugacities respectively. Finally, combining these equations, the permeability is given as a function of and : [166].

$$P_e = \frac{D_o c_s}{\Delta f} \ln \ln \left(1 + \frac{b\Delta f}{1 + bf_d} \right) \qquad (68)$$

1.2.2.12 DERJAGUIN-BROEKHOFF-DEBOER MODEL FOR PSD CALCULATING IN MESOPOR CARBONS

An improvement of the classical DBD theory for capillary condensation/evaporation in open-ended cylindrical capillaries was presented in Ref. [42]. Here, we reintroduce the main ideas of the DBD theory and present its extension for the capillary condensation/evaporation in spherical mesopores. It was previously shown that the experimental adsorption data for a reference flat silica surface can be properly described by using the disjoining pressure isotherm in the equation:

$$\Pi_1 \exp(-h/\lambda_1) + \Pi_2 \exp(-h/\lambda_2) = -(RT/v_m) \ln \ln (p/p_0) \qquad (69)$$

in whichandcharacterize the strength of the surface forces field, whereas the parameters and are responsible for the range of the structural forces action. Clearly, the first term dominates in thick ad layers, whereas the second term dominates in thin ad layers. All of the parameters appearing in Eq. (69) were tabulated previously for the adsorption of argon and nitrogen at their boiling points on the selected reference silica surface. The critical radius, at which a spontaneous capillary condensation occurs (spinodal condensation point), is closely related to the assumed pore geometry. For the wetting films formed on a concave surface of spherical pores, the following relationship is valid:

$$\Pi(h) = \Pi_1 \exp\left(-\frac{h}{\lambda_1}\right) + \Pi_2 \exp\left(-\frac{h}{\lambda_2}\right) + \frac{2\gamma(r_m)}{r-h} = -\left(\frac{RT}{v_m}\right)\ln\ln\left(\frac{p}{p_0}\right) \qquad (70)$$

As demonstrated previously, the surface tension of a liquid adsorbate depends on the meniscus radii , which seems to be particularly important for the pores at the borderline between micropores and mesopores. Similar to the work completed by Miyahara and co-workers, the GTKB was used in this work. The GTKB equation for the cylindrical interface (capillary condensation) can be written as follows:

$$\frac{\gamma(r_m)}{\gamma_\infty} = 1 - \frac{\delta}{r_m} \qquad (71)$$

In Eq. (71), denotes the surface tension of the bulk fluid, and is the displacement of the surface at zero mass density relative to the tension surface. The physical meaning of and its impact on the spinodal condensation point was presented previously. The stability condition of the wetting film was formulated earlier by Derjaguin et al. as. Obviously, Capillary Condensation/Evaporation in Spherical Cavities both the critical film thickness, , and the critical capillary radius, , corresponding to the film collapse, are determined from[42]

$$\frac{d\Pi(h)}{dh}\Big|_{h=h_{cr}, r=r_{cr}} = 0 \qquad (72)$$

The condition given by Eq. (72) determines the spontaneous spinodal condensation when the adsorbed film thickness becomes mechanically unstable.

A combination of Eqs. (69)–(72) gives the relation between both the critical film thickness and the critical capillary radius as a function of the relative pressure for the spherical pore geometry. The solution of this system of algebraic equations can be obtained by chord or other standard numerical procedures. The thickness of the adsorbed film in equilibrium with the meniscus for the spherical pore is given by:

$$RT\ln(p_0/p) = \Pi(h_e)v_m + \frac{2v_m\gamma_\infty\left(1-\dfrac{\delta}{r-h_e}\right)}{r-h_e} \qquad (73)$$

For constant surface tension, Eqs. (72) and (73) reduce to the classical DBD equations, which are described in the series of papers published by scientits. For application purposes, we can derive the analytical formulas for the calculation of the equilibrium transition in the considered spherical pore geometry according this equation. As we mentioned above, for the spherical pore, the system of equations describing the equilibrium desorption transition is defined by the following relation:

$$\theta_{exp}(h) = \int_{\dot{U}(D)} \theta_{loc}(D_{in},h)\chi(D_{in})dD_{in} \qquad (74)$$

Here,

$$\theta_{loc}(D_{in},h) = \left\{ \begin{array}{l} \dfrac{t}{t_{cr}}, t < t_{cr} \\ 1, elsewhere \end{array} \right\} \qquad (75)$$

represents the kernel of the theoretical isotherms generated from the DBD approach in pores of different diameters, , and is the normalized differential PSD function. As mentioned above, the kernel of the integral equation (given by Eq. (74)) was obtained assuming there is spinodal condensation in the spherical pores. By applying (as proposed by us previously) the ASA algorithm along with the stabilizing first-order Tihkonov's regularization term and regularization parameter was selected through a series of trials by an interactive judgment of the solution. During the regularization, some of the artificial shoulders and stairs were smoothed out.

1.2.2.13 DRS AND DI MODEL (DUBBINI-IZOTOVA, DUBBINI-RADUSKEVICH-STOEKLI)

The classical Dubinin–Radushkevich (DR) equation is commonly used in its linear form for analysis of active carbons and micropore size analysis that mention before. The DR equation was applied to the nitrogen experimental isotherm at 77 K using values. From this data, the micropore width (L) and following equations were used with the assumption of a slit-shaped pore:

$$\log\log(W) = \log\log(W_0) + M \times \log^2(P_0 / P) \qquad (76)$$

The micropore volume is calculated from the intercept of a log (W) vs plot, while the slope, M, of the best fit line is related to the adsorption energy, , as follows:

$$M = -2.303 \times (RT / E)^2 \qquad (77)$$

where is the ideal gas constant and is the adsorption temperature in Kelvin. It has also been shown that the parameter is related to the average micropore half-width by the equation:

$$x = \frac{\beta k}{E} \qquad (78)$$

The similarity coefficient is a shifting factor, which at a given temperature depends only on the adsorbate and is equal to , where is the adsorption energy of a reference vapor (typically benzene). For nitrogen at 77 K, is equal to a value of 0.33 and is a structural parameter that is equal to 13 nm kJ/mol for this set of materials. The micropore width, , is calculated as twice the micropore half-width, . The Dubinin–Radushkevich–Stoeckli (DRS) equation was used to determine the pore size distributions for several of the ACFs. Previously, Daley and co-workers showed a good correlation between this theory and direct measurement of the pore size distributions using STM.

The fitting of simulation data by the studied models was performed using the genetic algorithm of simulation. All results were described (in the whole pressure range) by a classical Dubinin–Astakhov adsorption isotherm equation, using the values of the affinity coefficient tabulated in this review paper. Moreover, we applied the Dubinin–Izotova model and the Dubinin– Raduskevich–Stoeckli equation (also in the whole pressure range) in the form:

$$N_{DRS} = \frac{N_{mDRS}}{(1+erf(\frac{x_0}{A\sqrt{2}}))\sqrt{1+2m\Delta^2 A_{pot}^2}} \times \exp\left[-\frac{A_{pot}^2 m x_0^2}{1+2m\Delta^2 A_{pot}^2}\right] \times \left[1+erf\left(\frac{x_0}{A\sqrt{2}\sqrt{1+2m\Delta^2 A_{pot}^2}}\right)\right] \quad (79)$$

where and are the values of adsorption and maximum adsorption, respectively,

is the adsorption potential, is a proportional coefficient (is assumed as equal to 12 (kJ nm mol^{-1})), is the similarity coefficient, erf is the error function, and are 'dispersion' and mean of Gaussian distribution, respectively. The pore size distribution was calculated using the correct normalization factor (i.e., from 0 up to):

$$\chi_{normDRS} = \frac{2}{\Delta\sqrt{2}\pi(1+erf\left(\frac{x_0}{\Delta\sqrt{2}}\right))} \quad (80)$$

Finally, the data were also described using the model proposed by Jaroniec and Choma, where

$$\chi_{JCh}(x) = \chi_{normJCh} \exp\left[-\rho\zeta x^n\right] \quad (81)$$

and

$$\chi_{normJCh} = \frac{n(\rho\zeta)^{\nu+1}}{\tilde{A}(\nu+1)} \quad (82)$$

where is the Euler gamma function, is constant equal to and are the parameters of Eqs. (80)–(82). The average micropore diameters from DI model were calculated using:

$$H_{eff,av,DI} = \frac{N_{m1}H_{eff,av1} + N_{m2}H_{eff,av2}}{N_{m1}+N_{m2}} \quad (83)$$

For the remaining models the average Micropore diameters were calculated from integration of the PSD curve.

1.2.2.14 AVERAGE MICROPORE DIAMETER (L_{AV}), CHARACTERIZATION ENERGY (E_0) CALCULATION AND ADSORPTION CALORIMETRIC DATA

Relation between average micropor diameter (L_{av}) and characterization energy (E_0), is presented by Eq. (84):

$$L_{av} = \frac{\kappa}{E_0} \tag{84}$$

where is the characteristic constant for a defined adsorbate/adsorbent pair in the micropore region. The value of this characteristic constant for benzene vapor on activated carbon is about 12 kJ nm mol^{-1}. It has been also assumed that this parameter is in a small degree dependent on the characteristic energy of adsorption:

$$L_{av} = \frac{13.028 - 1.53 \times 10^{-5} E_0^{3.5}}{E_0} \tag{85}$$

Next, others relations were developed since this turning point. The intensive investigations were done by McEnaney:

$$L_{av} = 6.6 - 1.79 \ln \ln \left(E_0 \right) \tag{86}$$

Stoeckli and co-workers model:

$$L_{av} = \frac{10.8}{E_0 - 11.4} \tag{87}$$

$$L_{av} = \left(\frac{30}{E_0} \right) + \left(\frac{5705}{E_0^3} \right) + 0.028 E_0 - 1.49 \tag{88}$$

Choma and Jaroniec model:

$$L_{av} = \left(\frac{10.416}{E_0} \right) + \left(\frac{13.404}{E_0^3} \right) + 0.008212 E_0 + 0.5114 \tag{89}$$

and Ohkubo et al.:

$$\ln \ln \left(p / p_s \right) = \frac{A_{HK}}{L - d} \left[\frac{B_{HK}}{\left(L - d/2 \right)^3} - \frac{C_{HK}}{\left(L - \frac{d}{2} \right)^9} - D_{HK} \right] \tag{90}$$

where is the sum of the diameter of an adsorbent atom and an adsorbate molecule (the remaining parameters of the above equation are defined in list of abbreviation). Finally, the following two equations were obtained:

$$\frac{L_{av(all)}}{d_A} = \frac{a_{1(all)}}{1 + b_{1(all)} \exp\left[-c_{1(all)}n\right]} + \frac{a_{2(all)} + b_{2(all)} \times n + c_{2(all)}n^2}{E_0} \tag{91}$$

not only for the micropore diameters (called the first range):

$$\frac{L_{av(mic)}}{d_A} = \left(\frac{a_{1(mic)}n}{b_{1(mic)} + n}\right) + \left(\frac{a_{2(mic)}n}{b_{2(mic)} + n}\right)^{E_0} \times E_0^{\left(\frac{a_{3(mic)}n}{b_{3(mic)} + n}\right)} \tag{92}$$

for calculations up to effective diameter equal to 2 nm (called the second range).

The additional details of all above-mentioned calculations were given previously, therefore they are omitted in the current study. Taking into account the assumptions made during the derivation of Eq. (90), the main condition, which should be fulfilled for the chosen molecules of adsorbates, is the absence of a dipole moment and a spherical-like structure. Therefore, with chosen N_2, Ar, CCl_4, and C_6H_6 for some of these adsorbates the Horvath–Kawazoe method was adopted. Moreover, these adsorbates have been widely applied in the investigation of the structural heterogeneity of microporous carbons. However, the choice of nitrogen at its liquid temperature as the probe molecule may not be suitable for very narrow pores such as those in carbon molecular sieves, where the activated diffusion effects might be important. These effects can be reduced by conducting the experiment at higher temperatures. It is thus useful to investigate the MSD obtained from adsorption of different adsorbates at temperature other than the boiling point of nitrogen at 77.5 K, For example, at near ambient temperatures. The temperatures chosen in the calculation were equal or very close to those applied in measurements where the investigated adsorbates are used for the determination of structural heterogeneity of carbons. Summing up, all the above results show that the average pore diameter is a function not only of E_0 but also of n. Then, using Eqs. (91) and/or (92), the average reduced effective diameter can be calculated and multiplied by the adsorbate diameter. The typical plots (the lines) for chosen adsorbates (C_6H_6 and N_2) and values of (1.50 and 3.25) are compared with relationships proposed by the other others works. It should be pointed out that the shape of this curves is similar to that observed for the empirical and/or semiempirical relationships. This procedure was also applied previously to experimental data of adsorption on different carbonaceous molecular sieves. The correlation between suggested and calculated using Eqs. (91) and (92) pore diameters is very good. On the other hand, we recently tried to answer the most general questions

[133]. In addition, to perform the thermodynamic verification for both samples studied in this paper, Eq. (93) is:

$$q^{diff} = \Delta G^{ads} - \Delta H^{vap} \tag{93}$$

whereand is the enthalpy of vaporization (equal with minus sign to the enthalpy of condensation,). Horvath and Kawazoe noticed the similarity of the data calculated based on Eq. (93) (with the "experimental" isosteric heat of adsorption obtained from isotherms measured at different temperatures. It should be pointed out that the HK model has been verified only for one set of experimental data and we do not find other cases in the literature. Therefore, the calorimetrically measured enthalpies of adsorption of C_6H_6 and CCl_4 are shown in Eqs. (94) and (95). Moreover, these data will be applied to calculate, using the standard method the enhancement of potential energy in micropores in comparison to the energy of adsorption on a "flat" surface. Knowing adsorption isotherms and the differential molar heats of adsorption, the differential molar entropies of adsorbed molecules () can be calculated by:

$$S^{diff} = S_g - \left(\frac{q^{diff}}{T}\right) - R\ln\ln\left(\frac{p}{p_0}\right) + R \tag{94}$$

where is the molar entropy of the gas at the temperature and is the standard state pressure. It is well known that different standard states can be chosen, and in our case, the gas at the standard pressure of = 101,325 Pa was applied. Also some researchers calculated the enthalpy and entropy of adsorption basing on the potential theory and applying the procedure described previously. Assuming the fulfillment of the main condition of the potential theory (first of all the temperature invariance condition that obtained:

$$q^{diff} = A_{pot} + \frac{\alpha T\Theta}{F\left(A_{pot}\right)} + L - RT = A_{pot} + \frac{\alpha T\beta E_0}{n}\left(\frac{A_{pot}}{\beta E_0}\right)^{1-n} + L - RT \tag{95}$$

1.2.2.15 BIMODAL PORE SIZE DISTRIBUTIONS FOR CARBON STRUCTURES

Very puzzling results are obtained by analyzing microporosity of various carbonaceous materials of different origin and/or treated thermally or chemically since the difference between the micropore size distribution curves are often insignificant. The number of peaks on the pore size distribution (PSD) curves and the ranges of their location as well as their shapes are very similar. Thus, the bimodality of PSD for many carbonaceous adsorbents is well known fact in the literature. This fact can be explained by conditions of carbon preparation; for example, by creation of the porous structure during activation process. It is amazing that different carbon materials obtained from different precursors have similar (still bimodal) porous structure with the gap between the peaks reflecting to the filling of primary and secondary micropores. Since the recent recipes for the preparation and manu-facturing of new carbonaceous adsorbents are still developed, the studies of the methods for porosity characterization should be improved and this improvement should enhance the method sensitivity. It is not surprising that the above topics attract a lot of attention. Moreover, a systematic numerical investigation of the effect of bimodality of PSD and its sensitivity on the reconstruction has not been published yet. Some researchers observed that the micropore size distributions determined from experimental isotherms (GCMC and DFT) usually show minima near two and three molecular diameters of the effective pore width (1 nm), re-gardless of the simulation method used. The proposed explanation was that this is a model induced artifact arising from the strong packing effects exhibited by a parallel wall model. Moreover, the inclusion of surface heterogeneity in the DFT model used to generate the local adsorption isotherms (while more realistic) did not change this observation significantly. Scientists argued that these artificial minima are primarily due to the homogeneous nature of the model (a sharp mono-layer formation occurring at approximately the same relative pressure in most of the pores, and the second layer existence in pores between 1.0 and 2.0 nm) that can accommodate four or more layers forming at the relative pressure of ca. 0.1.In the other words, since all theoretical isotherms in wide pores exhibit a monolayer step at about the same pressure, the contribution from pores that fill at this pres-sure would have to be reduced. This compensation effect is responsible for the observed minima on the pore size distribution curves, and also for the deviations between best fit and experimental isotherms. Similarly, less pronounced minima occur for pores, which fill at the relative pressures corresponding to the formation of the second, third, and higher layers. In addition, scientists investigated the ef-fect of heterogeneity of the pore walls (differences in the pore wall thicknesses) on the PSD shape. They pointed out that the influence of the pore wall thickness is significant, especially for lower values of relative pressures since the intermolec-ular potential is dominated by interactions with this wall. Consequently, the pore size distributions are shifted to smaller pores as the pore wall thickness decreases. However, the minima at 1and 2 nm were observed for all systems studied. The impact of the boundary values of nitrogen relative pressure (p/ps) was analyzed on the basis of the observed alterations in PSDs by many scientits.

They observed that the changes in microporosity and mesoporosity of acti-vated carbons can be described adequately only when the range of p/p_s is as wide as possible. The PSD curves can be broadened with shifted maxima especially for micropores and narrow mesopores when adsorption data start at relatively high p/ps value. However, the differences between the respective pore size distribution functions are insignificant. The influence of a random noise on the stability of the solution of the integral equation (calculation of PSD) was also studied on the basis of theoretically generated isotherms. For low and medium noise levels, the reproducibility of some perturbed isotherms was good. Summing up, the "true" and some perturbed generated isotherms (also experimental ones) contain full information about the assumed distribution (number of peaks, their location, area, etc.). Moreover, it was shown that an increase of the smoothing (regularization, the INTEG algorithm) parameter leads to a strong smoothness and disappearance of some (reasonable for experimental studies) peaks on the distribution curve (i.e., the PSD shape changes from initial polymodal distribution to much flatter monomodal one).It should be pointed out that similar results were obtained by others. In recent studies some systematic investigations of the influence of pore structure and the adsorbate–adsorbate and adsorbate–adsorbent parameters on the pore size distribution functions were performed. Thus, it was shown on the basis of experimental and theoretical data that the change in the shape and behavior of the local adsorption isotherm do not guarantee the differences between the evaluated PSDs except the strictly microporous adsorbents. All the observations suggested that for some parameters the larger pore diameters of the PSD function are, the smaller changes in the value of average pore diameter were observed. Moreover, the influence of the energetic heterogeneity on the structural parameters and quantities is rather insignificant; all calculations led to almost the same PSD curves and this similarity was observed for the adsorbents of different origin and possessing different pore structures. In conclusion, the following question arises: Why and when the gap between peaks of the differential pore size distribution curves can be related to the mechanism of the primary and secondary micropore filling? In order to answer this question we will use the experimental data published by some researchers showing the development of secondary porosity in a series of carbons. These data will be described by applying the proposed previously the ASA algorithm with the method of Nguyen and Do (ND). Next it will be shown that the behavior of the experimental systems studied represents a specific case, which can be, together with other systems, the subject of the proposed general analysis of the behavior of PSD curves. Therefore, by using carbon samples with varying porous structure it is possible to perform a systematic study of various situations related to the shape of the PSD curves, that is, the intensity of the both peaks, their mutual location and the vanishing of one of them. It is also possible to obtain the information how far the bimodality of those distributions is retained and reconstructed. Moreover, the problem in the similarity of the local adsorption isotherms generated for the range of pore widths corresponding to the gap between peaks will be discussed way, that is, the PSD curves (J(Heff)) using the ASA algorithm and the method proposed by Do and co-workers are calculated. It should be noted that 82 local isotherms generated for the same effective widths changing from 0.465 up to 233.9 nm were used. Additional details of the preparation of carbons and the procedure of the differential PSD calculations were given previously. Nearly all distribution curves show the existence of the bimodal porous structure. The location of the maximum of the first peak and its intensity are similar for all samples studied and only some changes in the width of this peak are observed.

1.2.2.16 COMPARISON BETWEEN DIFFERENT MATHEMATICAL MODELS FOR PSD CALCULATING AND ADVANTAGES AND DISADVANTAGES OF THESE MODELS

This chapter has reviewed modeling over the years and the progress that can be identified. It has been demonstrated that modelers, using modem computational systems, base their calculations on structure systems, usually around the graphitic microcrystallite. An objective of such modelers is the simulation of adsorption isotherms determined experimentally. Whether or not the structural models assumed for their work can be considered to be "realistic" is rarely a matter for discussion. Accordingly, the considerations of such modelers should not be adopted, uncritically, by those who have other interests in activated carbon. There is no unique structure within an activated carbon, which provides a specific isotherm, For example, the adsorption of benzene at 273 K. The isotherm is a description of the distribution of adsorption potentials throughout the carbon, this distribution following a normal or Gaussian distribution. If a structure is therefore devised whichpermits a continuous distribution of adsorption potentials, and this model predicts an experimental adsorption isotherm, this then is no guarantee that the structure of the model is correct. The wider experience of the carbon scientist, who relates the model to preparation methods and physical and chemical properties of the carbon, has to pronounce on the "reality" or acceptance value of the model. Unfortunately, the modeler appears not to consult the carbon chemist too much, and it is left to the carbon chemist to explain the limited acceptability of the adopted structures of the modeler. This approach of the modeler is still to be found in many publications. Different models rely on different assumptions in order to obtain relationships allowing the calculation of the main characteristics of adsorbent structural heterogeneity. The assumptions are summarized as follows:

(a) The shape of the pores is assumed to be slit like with effects of interconnectivity to be neglected.

(b) The molecule-surface interactions obey the equation of steel.

(c) The adsorbed phase can be considered as liquid-like layers between the two parallel walls of the pore, as a monolayer or as a double layer.

(d) The thickness of the adsorbed layer is determined by the distance between two parallel walls of the slit-like porosity, the density of the layers being constant between the layers and being equal to that of the bulk liquid.

(e) No gas-phase adsorbate exists within the volumes between the layers [1, 2, 133].

Earlier, Olivier [10] used GCMC and DFT as modeling methods based on slit-like pores with graphitic surfaces, as used by Kowalczyk et al. [15]. Olivier [10] concluded that the inclusion of surface heterogeneity into the model would have little effect on obtained simulations. Further, Olivier [10] considers that the use of alternative geometries maybe worth pursuing. The approach of is interesting in that several quite different approaches have been brought to the modeling of activated carbons. The concept of porosity not accessible to nitrogen at 77 Kis understood. A different concept, that the surface structure (in terms of pore density) is not representative of the bulk density of porosity, can be commented upon adversely. The suggestion that the fibrils of botanical structure of the olive stones have a role to play in maintaining the mechanical strength of the stones may need to be modified in the light of modem analyzes. The approach of this Section is limited because of confusion over what are acceptable surface area values and what are not acceptable surface area values. In addition, this section makes two assumptions which can be commented upon critically, namely that carbons can be categorized into well-defined groups according to their pore volumes and mean pore dimensions, and that it is adsorption on graphitic microcrystallites which controls the creation of an isotherm. Finally, the use of two mazes, two-dimensional and three-dimensional, designed as puzzle games provides helpful models to understand the complexity of the network of microporosity within a carbon. The simple ratio of carbon atoms to nitrogen molecules adsorbed as a monolayer capacity could be useful in assessments of how the microporosity fits into the carbon layer networks. This PSD calculating Models designed, to describe the microporous nature of carbons have been compared and contrasted leading to an assessment of the requirements of a comprehensive model to account for the properties of microporous systems. No comprehensive model has, as yet, been created. These of a model, based on a maze, provide insights. In the different adsorption processes, both in gas and liquid phase, the molecules or atoms adsorbable) are fixed (adsorbed) on the carbon (adsorbent) surface by physical interactions (electrostatic and dispersive forces) and/or chemical bonds. Therefore, a relatively large specific surface area is one of the most important properties that characterize carbon adsorbents. The surface of the activated carbons consists mainly of basal planes and the edges of the planes that form the edges of microcrystallites. Adsorption capacity related parameters are usually determined from gas adsorption measurements. The specific surface area is calculated by applying the Brunauer-Emmett-Teller (BET) equation to the isotherms generated during the adsorption process. The adsorption of N_2 at 77 K or CO_2 at 273 K are the most commonly used to produce these isotherms. The BET theory is basedupon the assumption that the monolayer is located on surface sites of uniform adsorption energy and multilayer build-up via a process analogous to the condensation of the liquid adsorbate. For convenience, the BET equation is normally expressed in the form which requires a linear relationship between p/p_0, and model parameters from which the monolayer capacity, nm (mmol g^{-1}), can be calculated. In activated carbons the range of linearity of the BET plot is severely restricted to the p/p_0 range of 0.05–0.20. The alternative form of linearization of the BET equation appears to be more convenient for a microporous solid since the choice of the appropriate experimental interval is free of ambiguity. The BET equation, however, is subject to various limitations when applied to microporous carbons. Thus, constrictions in the microporous network may cause molecular sieve effects and molecular shape selectivity [1, 2, 133].

Diffusion effects may also occur when using N_2 at 77 K as the adsorbate since at such low temperatures the kinetic energy may be insufficient to penetrate all the micropores. For this reason adsorption of CO_2 at higher temperatures (273 K) is also used. CO_2 and N_2 isotherms are complementary. Thus, whereas from the CO_2 isotherm micropores of up to approximately 10^{-9} m width can be measured, the N_2 can be used to test larger pores. Despite these limitations the BET surface area is the parameter most commonly used to characterize the specific surface area of carbon adsorbents. On the basis of volume-filling mechanism and thermodynamic considerations, Dubinin and Radushkevich found empirically that the characteristic curves obtained using the Potential Theory for adsorption on many microporous carbons could be linearized using the Dubinin-Radushkevich (DR) equation. For some microporous carbons the DR equation is linear over many orders of magnitude of pressure. For others, however, deviations from the DR equation are found. For such cases the Dubinin-Astakhov equation has been proposed in which the exponent of the DR equation is replaced by a third adjustable parameter, n, where $1 < n < 3$. Both the BET and the Dubinin models are widely thought to adequately describe the physical adsorption of gases on solid carbons. BET surface areas from many microporous carbons range from 500 to 1500 $m^2 \, g^{-1}$. However, values of up to 4000 $m^2 \, g^{-1}$ are found for some super activated carbons and these are unrealistically high. The relatively high values of the surface areas of activated carbons are mainly due to the contribution of the micropores and most of the adsorption takes place in these pores. At least 90–95% of the total surface area of an activated carbon may correspond to micropores. However, meso and macropores also play a very important role in any adsorption process since they serve as the passage through which the adsorbate reaches the micropores. Thus, the mesopores, which branch off from the macropores, serve as passages for the adsorptive to reach the micropores. In such mesopores capillary condensation may occur with the formation of a meniscus in the adsorbate. Although the surface area of the mesopores is relatively low in most activated carbons, some may have a well-developed mesoporosity (200 $m^2 \, g^{-1}$ or even more). In addition, depending on the size of the adsorbate molecules, especially in the case of some organic molecules of a large size, molecular sieve effects may occur either because the pore width is narrower than the molecules of the adsorbate or because the shape of the pores does not allow the molecules of the adsorbate to penetrate into the micropores. Thus, slit-shaped micropores formed by the spaces between the carbon layer planes are not accessible to molecules of a spherical geometry, which have a diameter larger than the pore width. This means that the specific surface area of a carbon is not necessarily proportional to the adsorption capacity of the activated carbon. Pore size distribution, therefore, is a factor that cannot be ignored. The suitability of a given activated carbon for a given application depends on the proportion of pores of a particular size. In general highly microporous carbons are preferred for the adsorption of gases and vapors and for the separation of gas molecules of different dimensions if the carbon possesses a suitable distribution of narrow size pores (molecular sieves) while well-developed meso and macroporosity is necessary for the adsorption of solutes from solutions [145].

It is the entrance dimension and shape, which controls the adsorption process, be hexagonal (an appropriate shape), or circular or square. Once the adsorbate molecule is through the pore entrance, then the characteristics of the adsorbent take over and the isotherm is created that is not enough, because the processes of physical and chemical activation have to be understood and to do this, requires three-dimensional models porosity in carbons. It would be an advantage to have some idea of the structure of this network include of carbon atoms, in three dimensions, in order to understand the extraction (gasification) process. There are four limitations of importance. First, this maze, of course, is in two dimensions; second, the lines of the maze are too orientated relative to an x–y axis. Such parallelism is unlikely to exist within a porous carbon; third, this labyrinth is best suited to a microporous carbon, only and not to microporous carbon fibers. Fourth, in such a model, rates of diffusion are likely to be too slow and hence there is a need to consider the location of mesoporosity. The inclusion of mesoporosity is another matter. Mesoporosity has to promote enhanced adsorption to the interior of the fiber. As a matter of scaling, although the models that mentioned before provide an impressive number of adsorption locations, it will require too much of such models. The human mind cannot cope with this necessity. But, apart from these limitations the similarities are relevant enough:

1. There is a continuous connection between the lines throughout the labyrinth. All carbon atoms form part of a continuous graphene sheet.
2. There is a continuous connection of the routes (spaces) of the labyrinth. Hence, all adsorption sites are available to the adsorbate molecules.
3. The widths of the routes (spaces) of the labyrinth are not constant. Some are narrower than other. This is a very relevant point as it demonstrates, very clearly, the range of porosities, within the definition of microporosity of <2.0 nm, this accounting for molecular sieving effects.
4. Some of the routes of the labyrinth are barely visible (being very close to each other). This offers the suggestion that it is representative of closed porosity, that is closed to everything except helium and lithium, noting, on the way, that the term closed porosity is an imprecise term, meaning porosity not accessible to a specifically defined adsorbate molecule.
5. The PSD calculating models makes the point that access to the interior of the labyrinth is available from all external surfaces.
6. A close inspection of the edges of the lines of labyrinth (surfaces of the carbon surfaces) indicates a lack of smoothness, the edges being rough. This point is of importance to carbon science because the surfaces of po-

rosity are of imperfect graphene-like layers, with surface irregularities and other defects.

7. This model also demonstrates that most of the continuous three-dimensional grapheme layer is able to act as an adsorbent surface making use of both sides. Rarely, do parts of the graphene layer come together (stack) in two or three layers. The range of such "different" sites is a feature of the models. Clearly, all the requirements for molecular-sieve properties, the dynamics and enthalpies of adsorption are present even though the detail is absent. The effect of increasing HTT would be to remove defects or irregularities of the carbon atom network resulting in decreased enthalpies of adsorption on more homogeneous surfaces. Such a porous system would respond to SAXS and SANS with the range of ring structures in the carbon atom network being sufficient to explain observed Raman spectra and electron spin resonance (ESR) during the carbonization process [1, 133].

There are many methods for calculation of pore size distributions (PSDs) and most of them are potentially applicable for nanoporous carbons. PSDs for nanoporous carbons are usually evaluated using methods based on either the Kelvin equation or the Horvath-Kawazoe method and its modifications. The first group includes the models of Barrett, Joyner and Halenda (BJH) method, Cranston and Inkley (CI), Dollimore and Heal (DH), and Broekhoff and de Boer (BdB). Although the BJH, CI, and DH methods are often considered as appreciably different, all of them are based on the general concept of the algorithm outlined in the BJH work. To implement the algorithms proposed in these three methods, the knowledge of the relation between the pore size and capillary condensation or evaporation pressure and the t-curve is required, and a choice needs to be made which branch of the isotherm is appropriate for the PSD calculation. The original BJH, CI, and DH models are not fully consistent as far as the selection of these relations and the choice of the branch of the isotherm are concerned. These inconsistencies are capable of affecting the results of calculations much more than the minor differences in the algorithms, being most likely responsible for claims that these three methods differ substantially. The BJH, CI and DH methods assume the same general picture of the adsorption desorption process. Adsorption in mesopores of a given size is pictured as the multilayer adsorption followed by capillary condensation (filling of the pore core, that is, the space that is unoccupied by the multilayer film on the pore walls) at a relative pressure determined by the pore diameter. The description is pictured as capillary evaporation (emptying of the pore core with retention of the multilayer film) at a relative pressure related to the pore diameter followed by thinning of the multilayer. Because the concept underlying the BJH, CI, and DH models appears to be correct, it is important to:

(i) Establish an accurate relation between the pore size and capillary condensation or evaporation pressure;

(ii) Determine to correct t-curve;

(iii) Verify whether adsorption or desorption, or both branches of the isotherms, are suitable for the accurate pore size assessment.

This would allow performing accurate PSD calculations using these simple algorithms. Theoretical considerations, non-local density functional theory (NLDFT) calculations computer simulations and studies of the model adsorbents strongly suggested that the Kelvin equation commonly used to provide a relation between the capillary condensation or evaporation pressure and the pore size calculating. Porosity of nanoporous carbonaceous materials is usually analyzed on the basis of nitrogen adsorption isotherms, which reflect the gradual formation of a multilayer film on the pore walls followed by capillary condensation in the unfilled pore interior. The pressure dependence of the film thickness is affected by the adsorbent surface. Hence, an accurate estimation of the pore size distribution requires a correction for the thickness of the film formed on the pore walls. The latter (so-called t-curve) is determined on the basis of adsorption isotherms on non-porous or macroporous adsorbents of the surface properties analogous to those for the adsorbent studied. Modified non-local density functional theory (MDFT) has been shown to provide an excellent description of the physical adsorption of nitrogen or argon on the energetically uniform surface of graphite. This and other formulations of DFT have been used to model adsorption in narrow slit pores and provide the basis for a method of estimating pore size distribution from experimental isotherms [1, 2, 145].

Important parameters that greatly affect the adsorption performance of a porous carbonaceous adsorbent are porosity and pore structure. Consequently, the determination of pore size distribution (PSD) of coal-based adsorbents is of particular interest. For this purpose, various methods have been proposed to study the structure of porous adsorbents. A direct but cumbersome experimental technique for the determination of PSD is to measure the saturated amount of adsorbed probe molecules, which have different dimensions. However, there is uncertainty about this method because of networking effects of some adsorbents including activated carbons. Other experimental techniques that usually implement for characterizing the pore structure of porous materials are mercury porosimetry, XRD or SAXS, and immersion calorimetry. A large number of simple and sophisticated models have been presented to obtain a relay estimation of PSD of porous adsorbents. Relatively simple but restricted applicable models such as Barret, Joyner and Halenda (BJH), Dollimore and Heal (DH), Mikhail et al. (MP), Horvath and Kawazoe (HK), Jaroniec and Choma (JC), Wojsz and Rozwadowski (WR), Kruk-Jaroniec-Sayari (KJS), and Nguyen and Do (ND) were presented from 1951 to 1999 by various researchers for the prediction of PSD from the adsorption isotherms. For example, the BJH method, which is usually recommended, for mesoporous materials is in error even in large pores with dimension of 20 nm. The main criticism of the MP method, in addition to the uncertainty regarding the multilayer adsorption mechanism in micropores, is that we should have a judicious choice of the reference isotherm. HK model was developed for calculating micropore size distribution of slit-shaped pore; however, the HK method suffers from the idealization of the micropore filling process. Extension of this theory for cylindrical and spherical pores was made by Saito and Foley and Cheng and Ralph. By applying some modifications on the HK theory, some improved models for calculating PSD of porous adsorbents have been presented. Many scientists extended the Nguyen and Do method for the determination of the bimodal PSD of various carbonaceous materials from a variety of synthetic and experimental data. The pore range of applicability of this model besides other limitations of ND method is its main constraint [1, 2, 133].

Many researchers determined the PSD of activated carbons based on liquid chromatography (LC). Choices of suitable solvent and pore range of applicability of this method are two main problems that restrict its general applicability. More sophisticated methods such as molecular dynamics (MD), Monte Carlo simulation, Grand Canonical Monte Carlo simulations (GCMS), and density functional theory (DFT) are theoretically capable of describing adsorption in the pore system. The advantages of these methods are that they can apply on wide range of pores width. But, they are relatively complicated and provide accurate PSD estimation based on just some adsorbates with specified shapes. Requiring extensive computation time and significantly different idealized conditions from a real situation are other drawbacks of such methods. Researchers used a new approach based on Monte Carlo integration to derive pore size and its volume distribution for porous solids having known configuration of solid atoms. They applied the proposed method to a wide range of commonly used porous solids. However, this method also seems to have limitations of Monte Carlo simulation in addition to requiring the information of solid atomistic configuration. Recently, some researchers proposed a multiscale approach based on Grand Canonical Monte Carlo (GCMC) to predict the adsorption isotherms and PSD of porous solids. However, the proposed methodology needs further improvement to provide high accuracy predictions besides mass of computations that is needed. Kind of PSD assumes in DA equation itself, is important in design of PSD modeling. Does this distribution change if the parameters of the DA isotherm change (especially n). Some researchers studied the simulated nitrogen adsorption isotherms in heterogeneous carbons and assumed Gaussian distribution of pores. It was shown that the simulated isotherms can be fitted by typical DR, although not in the whole range of relative pressures. Two groups of low-temperature (= 77.5 K) N_2 adsorption isotherms were generated, and studied the influence of n at constant E_0 and, to the contrary, the effect of E_0 at constant n. The obtained curves were converted into high-resolution -plots in order to explain the mechanism of adsorption. Moreover, the new algorithm (called the adsorption stochastic algorithm (ASA) was used to solve the problem of fitting the local adsorption isotherms of the Nguyen and

Do method to experimental data. The obtained results show that the DA equation generates isotherms describing almost a homogeneous structure of pores and/or a bimodal heterogeneous structure. Corresponding PSDs indicate the presence of homogeneous porosity, primary and/or secondary micropore filling, or both. The parameter of the DA equation is responsible not only for the homogeneity of pores (the deviation of pores from average size) but for the adsorption mechanism in micropores. In other words, lowering n leads to the change in this mechanism from primary to simultaneous primary and secondary micropore filling. Taking all obtained results into account suggests that the DA equation is probably the most universal description of adsorption in micropores. All theoretical models (as, for example, considered in the current study: HK, ND, and DFT, etc.) are connected with their own specific assumptions of the description of the porous structure and/ or mechanism of adsorption. Of course, those postulations can significantly influence obtained results, PSD curves. The modeling of local and/or global adsorption isotherms for different pore widths using DFT theory, Monte Carlo simulations, and the ND model leads to results that should be, for some cases, treated with caution. The main simplifications of those theories are, for example, the neglect of the pore connectivity, ignoring of different thickness of carbon microcrystallites forming micropores, or existence of various surface groups on the surface of activated carbons. Moreover, it is very difficult to find papers where authors obtained satisfactory results (using the abovementioned theories) describing simultaneously the experimental adsorption isotherm, adsorption enthalpy, and entropy (or heat capacity) for adsorption in microporous carbons around room temperature. Although all simulation and modeling of carbons is very interesting and sometimes spectacular, it should be remembered that small changes in the values of fundamental parameters taken as constants in calculations can lead to drastic changes in the results obtained. In this review, the results of this type of the calculation are speculative as long as a satisfactory model of the structure of carbons is not evaluated. It should be pointed out that very complicated models of the structure of a microporous activated carbon are sometimes considered in some advanced numerical and simulation calculations (for example, the RMC method, where the surface sites have been added at random points on the edges of the graphene microcrystals characterizing by differing size and shape) in order to describe the "real" structure of activated carbons. However, taking into account this complex structure is connected with considerable extension of the time of calculations. On the other hand, very puzzling results are obtained from the description of the micropore structure of various carbonaceous materials (the different origin and thermal or chemical treatment)) for the reason that the differences between the micropore size distribution plots are insignificant. For example, Ismadji and Bhatia published the results of microporosity determination from DFT method for three carbons, that the number of peaks on PSD curves and the ranges of their location, as well as the shapes, are very similar. In our opinion this behavior of

PSDs is very surprising and can be caused by the low sensitivity and the simplifications of the mentioned above methods [145].

The advantage of these models is that they are experimentally convenient and
do not need complicated PSD calculations. In 1998, Bhatia successfully applied
the combination of finite element collocation technique with regularization method to extract various double peak PSDs from synthetic isotherm data points con-
taminated with 1% normally distributed random errors using DR isotherm. They
applied the constraint of non- negativity of solutions by simply using a Newton-
Raphson technique. Although they reported that the method is stable over a wide
range of values of the regularization parameter, the application of non- negativity constraint usually provides unrealistic solutions. Some researchers proposed a
new method based on the modification of DR equation by introducing adsorption
density and correlating between the pore filling pressure and critical pore size
for nitrogen adsorption at 77 K. The results are found comparable with other
popular PSD methods such as MP, JC, HK, and DFT. In addition to uncertainty
about the general performance of this model as a result of some assumptions in
the model derivation, it is relatively complicated and the procedure for obtaining
PSD is cumbersome. The average diameter of the mesopores is usually calculated
from the nitrogen adsorption data using Kelvin equation. Recently, Shahsavand
and Shahrak proposed two new algorithms (SHN1, SHN2) for reliable extraction of PSD from adsorption and condensation branch of isotherms. According
to their results, it seems that these two methods have also some limitation similar
to the other previous ones. The regularization technique that used for obtaining
optimum value of regularization parameter is challengeable in these models. In
addition, the basis of these methods is Kelvin equation, whichunlikely provides
reliable PSD for microporous solids. Although much has been done to address
the PSD of porous adsorbents, up to now, no general reliable theory is available
leading to the conclusion that for microporous carbons the extensive investigation should still continue. In the recent study works, have tried as a novel work
to extend the analysis on PSD of porous carbon nano adsorbents by investigating
the effects of different parameters on it. Some well-known models based on Dubinin's method namely, Dubinin-Stoeckli (DS) and Stoeckli models and etc., were
used to investigate the effects of these parameters on the porosity of AC samples,
and the results were compared with the two widely used methods of Horvath-
Kawazoe (HK) and improved Horvath-Kawazoe (IHK) for the determination of
PSD of micro or mesoporous solids. These models were derived based on benzene as a reference adsorbate, or N_2 or Ar, because this gases provides more accurate estimations than other adsorbates. On this basis, adsorption isotherm data
of benzene at 30°Cor N_2in 77k or Ar in 87 Kwere used to determine the PSD of
each carbon samples. Table 1.5 summarizes novel models for characterization of
pores and determines the PSD in carbon materials [1, 2, 133].

TABLE 1.5 Techniques for Characterization of Pores in Carbon Materials.

Characterization Technique	Comments (advantages and disadvantages)
Adsorption/desorption of N_2 gas at 77K	
BET method	Give overall surface area (SS).
Alpha plot	Give microporous and external SSs separately.
	Give micropore volume.
BJH method	Differentiate microporous and mesoporous SSs and volumes.
	Give pore size distribution in mesopore range.
DFT method	Give pore size distribution in a wide range of size.
HK method etc.	Give pore size distribution
Adsorption/desorption isotherm of various gases (H_2, He, CO_2, CO, etc.)	Give the information of molecular sieving performance.
X-ray small-angle scattering	Detect micropores, both open and closed pores
Transmission electron microscopy	Detect nano-sized pores, even less than 0.4 nm size.
	Give localized of information, need statistical analysis of data
Scanning tunneling microscopy	Detect only pore entrances on the surface.
	Give morphological information of the pore entrance.
	Need statistical analysis with criteria.
Scanning electron microscopy	Detect only macropores.
Mercury porosimetry	Detect mostly macropores
	Difficult to apply for fragile materials

    BET and the Dubinin models (DRS, DA) beside HK and DFT models are widely thought to adequately describe the adsorption process of carbon nano adsorbents. For convenience, the BET equation is normally expressed in the form, which requires a linear relationship between p/p_0, and model parameters from which the monolayer capacity, and adsorption parameters can be calculated; in addition this method is more accessible. However, by compare of different model and simulation methods, this model is subject to various limitations applied to microporous carbons to calculating PSD in CNT-Textile composite, in our PhD study case in future.

1.3 SIMULATION METHODS FOR MODELING PROBLEMS: EXPECTATIONS AND ADVANTAGES

Simulation is the imitation of the operation of a real world or system over time to develop a set of assumptions of mathematical, logical and symbolic relationship between the entities of interest of the system, to estimate the measures of performance of the system with the simulation generated data. Simulation modeling can be used as an analysis tool for predicting the effects of changes to existing systems as a design tool to predict the performance of new systems. Simulation enables the study of, and experimentation with, the internal interactions of a complex system, or of a subsystem within a complex system. Informational, organizational, and environmental changes can be simulated, and the effect of these alterations on the model's behavior can be observed. The knowledge gained in designing a simulation model maybe of great value toward suggesting improvement in the system under investigation. By changing simulation inputs and observing the resulting outputs, valuable insight maybe obtained in to which variables are most important and how variables interact. Simulation can be used as a pedagogical device to reinforce analytic solution methodologies. Simulation can be used to experiment with new designs or policies prior to implementation, so as to prepare for what may happen. Simulation can be used to verify analytic solutions. By simulating different capabilities for a machine, requirements can be determined. Simulation is not appropriate in some cases include:

- When the problem can be solved using common sense and the problem can be solved analytically.
- When it is easier to perform direct experiments and the simulation costs exceed the savings.
- When the resources or time are not available and system behavior is too complex or can't be defined.
- When there isn't the ability to verify and validate the model.
- Simulation is the appropriate tool in some cases include:
- Simulation can be used to experiment with new designs or policies prior to implementation, so as to prepare for what may happen.
- Simulation can be used to verify analytic solutions and by simulating different capabilities for a machine, requirements can be determined.
- Advantages of simulation include:

- New polices, operating procedures, decision rules, information flows, organizational procedures, and soon can be explored without disrupting ongoing operations of the real system.
- New hardware designs, physical layouts, transportation systems, and soon, can be tested without committing resources for their acquisition.
- Hypotheses about how or why certain phenomena occur can be tested for feasibility.
- Insight can be obtained about the interaction of variables.
- Insight can be obtained about the importance of variables to the performance of the system.
- A simulation study can help in understanding how the system operates rather than how individuals think the system operates.
- "What-if" questions can be answered. This is particularly useful in the design of new system.
- The two main families of simulation technique are molecular dynamics (MD) and Monte Carlo (MC); additionally, there is a whole range of hybrid techniques which combine features from both.

For example, in the recent study works about PSD calculating simulated samples, the average of numbers adsorbed molecules fluctuates during the simulation calculating of for a range of chemical potentials enables the adsorption isotherm to be constructed. The walls of the slit pores lie in the $x–y$ plane. Normal periodic boundary conditions, together with the minimum image convention, are applied in these two directions. For low pressures, P/P_0 that smaller than 0.02, the length of the simulation cell in the two directions parallel to the walls was maintained at 10 nm for each of the pore widths studied, to maintain a sufficient number of adsorbed molecules. For higher pressures where more water molecules were present, the minimum cell length in the x and y directions was 4 nm. The average number of water molecules in the simulation cell varied from a few molecules at the lowest pressures to a few hundred or thousands of molecules when the pores were full; filled pores contained about 320 molecules for a pore width of 0.79 nm, 460 at 0.99 nm, 830 at 1.69 nm. Calculations were carried out on the Cornell Theory Center IBM SP². In determining the adsorption isotherm, commenced with the cell empty; a value of the fugacity corresponding to a low pressure was chosen and the average adsorption determined from the simulation. The final configuration generated at each stage was used as the starting point for simulations at higher fugacities. The pressure of the bulk gas corresponding to a given chemical potential was determined from the ideal gas equation of state. Gas phase densities corresponding to the range of chemical potentials studied were determined by carrying out simulations of the bulk gas. These were found to agree with those calculated from the ideal gas equation within the estimated errors of the simulations.

1.3.1   MONTE CARLO SIMULATION, APPLICATION, IMPORTANCE AND ITS AIMS

In a Monte Carlo simulation we attempt to follow the 'time dependence' of a model for which change, or growth, does not proceed in some rigorously pre-denned fashion (according to Newton's equations of motion) but rather in a stochastic manner which depends on a sequence of random numbers which is generated during the simulation method, in a simulation cell that design for special purposes. With a second, different sequence of random numbers the simulation will not give identical results but will yield values, which agree with those, obtained from the first sequence to within some 'statistical error.' A very large number of different problems fall into this category: in percolation an empty lattice is gradually filled with particles by placing a particle on the lattice randomly with each 'tick of the clock.' Considering problems of statistical mechanics, we may be attempting to sample a region of phase space in order to estimate certain properties of the model, although we may not be moving in phase space along the same path, which an exact solution to the time dependence of the model would yield. Remember that the task of equilibrium statistical mechanics is to calculate thermal averages of (interacting) many-particle systems. Monte Carlo simulations can do that, taking proper account of statistical errors and their effects in such systems. Many of these models will be discussed in more detail in later chapters so we shall not provide further details here. Since the accuracy of a Monte Carlo estimate depends upon the thoroughness with which phase space is probed, improvement may be obtained by simply running the calculation a little longer to increase the number of samples. Unlike in the application of many analytic techniques (perturbation theory for which the extension to higher order may be prohibitively difficult), the improvement of the accuracy of Monte Carlo results is possible not just in principle but also in practice.

The method of Monte Carlo simulation has proved very useful for studying the thermodynamic properties of model systems with moderately many degrees of freedom. The idea is to sample the system's phase space stochastically, using a computer to generate a series of random configurations. We take the phase space to consist of N discrete states (with label i), though the method applies equally to continuous systems. Often only a tiny fraction of the phase space (the part at low energy) is relevant to the properties being studied, due to the strong variation of the Boltzmann equation) in the canonical ensemble (CE). It is then helpful to sample in an ensemble (with relative weights w_i and absolute probabilities $p_i = w_i / P_j w_j$), which is concentrated on this region of phase space. The Metropolis algorithm samples directly in the CE, and is good at determines many physical properties [12, 17, 29].

The price to be paid for this is that successive configurations are not independent (typically they have a single microscopic difference), but instead form a Markov chain with some equilibration time $teq(w_i)$. We may distinguish two important characteristics of a Monte Carlo simulation: its ergodicity (measured by $teq(w_i)$) and its pertinence (measured by $Ns(w_i; I)$, the average number of independent samples needed to obtain the information I that we seek). We should choose w_i so as to minimize the total number of configurations that need to be generated, which is proportional to $teq(w_i)Ns(w_i; I)$. It is easy to specify an ensemble, which would yield the sought information if independent samples could be drawn from it, but an ensemble with too much weight at low energies may become fragmented into "pools" at the bottoms of "valleys" of the energy function, and so have a large equilibration time. For example, it is well known that at low temperatures the Metropolis algorithm can get stuck in ordered or glassy phases. Ergodicity may be improved by sampling instead in a non-physical ensemble with a broad energy distribution, which allows the valleys to be connected by paths passing through higher energies. A weight assignment leading to such a distribution cannot in general be written as an explicit function of energy alone; rather it is an algorithm's purpose to find this assignment, which then tells us about the density of states (E). This reversal (starting with the distribution and finding the weights) of the usual Monte Carlo process can be achieved using a series of normal simulations, adjusting the weight w_i after each run so that the resulting energy distribution $w_i(E)$ converges to the desired one. The last application reported here is a simulation of a regular system with frustration, the triangular anti ferromagnetic, on a 48×48 parallelogram with periodic boundary conditions. Using 5 runs of 7.4×105 sweeps, we obtained a ground state entropy of0.32320, with a variance of 0.00015, which is consistent with the exact bulk value $\simeq 0.32307$.As computers have improved in capability, the simulation of large statistical systems governed by known Hamiltonians has become an important tool of the theoretical physicist. Applications range from studies of phase transitions in condensed matter to calculation of hadronic properties via lattice gauge theory. Most of these simulations rely on adaptations of the algorithm of Metropolis. This generates a sequence of configurations via a Markovian process such that ultimately the probability of encountering any given configuration in the sequence is proportional to the Boltzmann weight function, $S(C)$, where is the energy for a statistical mechanics problem or the action for a quantum field theory simulation. Thus one obtains a sample of configurations, which dominate the partition function sum or path integral. An alternative simulation technique is the molecular dynamics or microcanonical method. This begins with a set of equations for a dynamical evolution, which conserves the total energy. Upon numerical integration the system will flow through phase space in a hopefully ergodic manner (Indeed, non-ergodic behavior would represent a fascinating exception to the generic case.). Such a program does not explicitly depend on an inverse temperature, which is determined dynamically by measuring, say, the average kinetic energy, which by equipartition should be IkT per degree of freedom. Note that the conventional microcanonical simulations make no use of random numbers, which are effectively generated by the complexity of the system. This technique has recently been applied to lattice gauge theory. Such equation gives another microcanonical formulation, which was discussed in the context of continuum field theory by Scientists.

Monte Carlo methods are used as computational tools in many areas of chemical physics. Although this technique has been largely associated with obtaining static, or equilibrium properties of model systems, Monte Carlo methods may also be used to study dynamical phenomena. Often, the dynamics and cooperatively leading to certain structural or configurational properties of matter are not completely amenable to a macroscopic continuum description. On the other hand, molecular dynamics simulations describing the trajectories of individual atoms or molecules on potential energy hyper surfaces are not computationally capable of probing large systems of interacting particles at long times. Thus, in a dynamical capacity, Monte Carlo methods are capable of bridging the ostensibly large gap existing between these two well-established dynamical approaches, since the "dynamics" of individual atoms and molecules are modeled in this technique, but only in a coarse-grained way representing average features which would arise from a lower-level result. The application of the Monte Carlo method to the study of dynamical phenomena requires a self-consistent dynamical interpretation of the technique and a set of criteria under which this interpretation may be practically extended. In recent publications, certain inconsistencies have been identified which arise when the dynamical interpretation of the Monte Carlo method is loosely applied. These studies have emphasized that, unlike static properties, which must be identical for systems having identical model Hamiltonians, dynamical properties are sensitive to the manner in which the time series of events characterizing the evolution of a system is constructed. In particular, Monte Carlo studies comparing dynamical properties simulated away from thermal equilibrium have revealed differences among various sampling algorithms. These studies have underscored the importance of using a Monte Carlo sampling procedure in which transition probabilities are based on a reasonable dynamical model of a particular physical phenomenon under consideration, in addition to satisfying the usual criteria for thermal equilibrium. Unless transition probabilities can be formulated in this way, a relationship between Monte Carlo time and real time cannot be clearly demonstrated. In many Monte Carlo studies of time-dependent phenomena, results are reported in terms of integral Monte Carlo steps, which obfuscate a definitive role of time. Ambiguities surrounding the relationship of Monte Carlo time to real time preclude rigorous comparison of simulated results to theory and experiment, needlessly restricting the technique. Within the past few years, the idea that Monte Carlo methods can be used to simulate the Poisson process has been advanced in a few publications and some Monte Carlo algorithms, which are implicitly based on this assumption, have been used. An attractive prospect, since within the theory of Poisson processes the relationship between Monte Carlo time and real time can be clearly established.

Early use of Monte Carlo techniques was made for the quantitative evaluation of fault trees. While some effort has continued in the use of purely Monte Carlo methods, they have largely been supplanted deterministic techniques often referred to as Kinetic Tree methods. Two limitations, however, present themselves in the use of Kinetic Tree methods. First, in Kinetic Tree methods the reliability characteristics of each component are modeled separately. To evaluate the fault tree by combining component failure probabilities, the components are assumed to have independently of one another. In fact, dependencies often arise from common mode failures, from the increased stress in partially disabled systems, and from a variety of errors in testing, maintenance and repair. Due to this limitation of the Kinetic Tree formulation there is increasing use of Markov models for reliability analysis, for with such models quite general dependencies between components may be treated. For systems with more than a few components, however, Markov analysis by deterministic means becomes a prodigious task. For even while innovative methods have been employed to reduce the complexity of the computations, the fact remains that one must solve a set of 21 coupled first-order differential equations, thus even a system with only ten components will result in a system of over one thousand coupled equations with a transition matrix with over a million elements. Moreover, if some of the components are repairable, the equations are likely to be quite stiff, requiring that very small time steps be used in the numerical integration. A second limitation on Kinetic Tree methods is a result of the lack of precision to which the component failure and repair rates are normally known. Invariably this is accomplished by Monte Carlo sampling of the failure rate data using log-normal or other distributions. The fault tree is evaluated deterministically with data from each data sampling, and the mean, variance and other characteristics of the system are estimated. A similar procedure is also applied to Markov models, requiring that the solution of the coupled set of differential equations be repeated thousands of times. What follows is the formulation of a class of Monte Carlo methods, which provides a natural framework for the treatment of both component dependencies and data uncertainties. Some researchers formulate Monte Carlo simulation of the unreliability of systems with repairable components within the framework of a Markov process. This approach retains the power of deterministic Markov methods in modeling component dependencies that would not be possible if direct Monte Carlo simulation were to be carried out. At the same time the Monte Carlo simulation requires very little computer memory. Variance reduction techniques, similar to those that have been highly developed for neutral particle transport calculations, are applied to greatly increase the computational efficiency of Monte Carlo reliability calculations. Monte Carlo formulation is generalized to include probability distributions that represent the uncertainty in the component failure and repair rate data. The variance in the result is then due to two causes: the finite number of random walk simulations, and the uncertainty in the data. A batching technique is introduced and is shown to further reduce that part of the variance due to the finite number of random walks without a commensurate increase in computing effort. The Markov Monte Carlo formulation was extended to problems with some data uncertainties, and a batching technique was shown to lead to further improvements in the figure of merit. While we have not had the opportunity to make numerical comparisons between Markov Monte Carlo and Kinetic Tree methods, which use Monte Carlo data sampling, an observation seems in order. For equal data sampling one would equate the number of Markov Monte Carlo batches to the number of Kinetic Tree trials. If one then chose the batch size just large enough so that the random walk variance could be ignored relative to the variance due to data uncertainty, a fair comparison of computational efficiency would be the Monte Carlo time per batch versus the Kinetic Tree time per trial. This, of course, assumes that the problem is chosen in which component dependencies do not rule out the use of Kinetic Tree methods.

Within the contents of this book we have attempted to elucidate the essential features of Monte Carlo simulations and their application to problems in statistical physics. We have attempted to give the reader practical advice as well as to present theoretically based background for the methodology of the simulations as well as the tools of analysis. New Monte Carlo methods will be devised and will be used with more powerful computers, but we believe that the advice given to the reader that will remain valid. In general terms we can expect that progress in Monte Carlo studies in the future will take place along two different routes. First, there will be a continued advancement towards ultra high-resolution studies of relatively simple models in which critical temperatures and exponents, phase boundaries, etc. will be examined with increasing precision and accuracy. As a consequence, high numerical resolution as well as the physical interpretation of simulational results may well provide hints to the theorist who is interested in analytic investigation. On the other hand, we expect that there will be a tendency to increase the examination of much more complicated models, which provide a better approximation to physical materials. As the general area of materials science blossoms, we anticipate that Monte Carlo methods will be used to probe the often-complex behavior of real materials. This is a challenge indeed, since there are usually phenomena, which are occurring at different length and time scales. As a result, it will not be surprising if multi scale methods are developed and Monte Carlo methods will be used within multiple regions of length and time scales. We encourage the reader to think of new problems which are amenable to Monte Carlo simulation but which have not yet been approached with this method. Lastly, it is likely that an enhanced understanding of the significance of numerical results can be obtained using techniques of scientific visualization. The general trend in Monte Carlo simulations is to ever-larger systems studied for longer and longer times. The mere interpretation of the data is becoming a problem of increasing magnitude, and visual techniques for probing the system (again over different scales of time and length) must be developed. Coarse-graining techniques can be used to clarify features of the results, which are not immediately obvious from inspection of columns of numbers. 'Windows' of various sizes can be used to scan the system looking for patterns, which develop in both space and time; and the development of such methods may well profit from interaction with computer science. Clearly improved computer performance is moving swiftly in the direction of parallel computing. Because of the inherent complexity of message passing, it is likely that we shall see the development of hybrid computers in which large arrays of symmetric (shared memory) multiprocessors appear. (Until much higher speeds are achieved on the Internet, it is unlikely that non-local assemblies of machines will prove useful for the majority of Monte Carlo simulations.) We must continue to examine the algorithms and codes, which are used for Monte Carlo simulations to insure that they remain well suited to the available computational resources. We strongly believe that the utility of Monte Carlo simulations will continue to grow quite rapidly, but the directions may not always be predictable. We hope that the material in this book will prove useful to the reader who wanders into unfamiliar scientific territory and must be able to create new tools instead of merely copying those that can be found in many places in the literature.

1.3.1.1 RANGE OF PROBLEMS CAN WE SOLVE WITH MONTE CARLO SIMULATION

The range of different physical phenomena, which can be explored using Monte Carlo methods, is exceedingly broad. Models, which either naturally or through approximation can be discretized, can be considered. The motion of individual atoms may be examined directly; for example, in a binary (AB) metallic alloy where one is interested in the diffusion or un mixing kinetics (if the alloy was prepared in a thermodynamically unstable state) the random hopping of atoms to neighboring sites can be modeled directly. This problem is complicated because the jump rates of the different atoms depend on the locally differing environment. Equilibrium properties of systems of interacting atoms have been extensively studied as have a wide range of models for simple and complex fluids, magnetic materials, metallic alloys, adsorbed surface layers, etc. More recently polymer models have been studied with increasing frequency; note that the simplest model of a flexible polymer is a random walk, an object that is well suited for Monte Carlo simulation. Furthermore, some of the most significant advances in understanding the theory of elementary particles have been made using Monte Carlo simulations of lattice gauge models [131].

1.3.1.2 THE PROBLEMS OF MONTE CARLO SIMULATION

1.3.1.2.1. LIMITED COMPUTER TIME AND MEMORY

Because of limits on computer speed there are some problems, which are inherently not suited to computer simulation, at this time. A simulation, which requires years of CPU time on whatever machine is available, is simply impractical. Similarly a calculation, which requires memory, which far exceeds that, which is available, can be carried out only by using very sophisticated programming techniques, which slow down running speeds and greatly increase the probability of errors. It is therefore important that the user first consider the requirements of both memory and CPU time before embarking on a project to ascertain whether or not there is a realistic possibility of obtaining the resources to SIM [136].

1.3.1.2.2   STATISTICAL AND OTHER ERRORS

Assuming that the project can be done, there are still potential sources of error, which must be considered. These difficulties will arise in many different situations with different algorithms so we wish to mention them briefly at this time without reference to any specific simulation approach. All computers operate with limited word length and hence limited precision for numerical values of any variable. Truncation and round-off errors may in some cases lead to serious problems. In addition there are statistical errors, which mention before. What difficulties will we encounter? An inherent feature of the simulation algorithm due to the finite number of members in the 'statistical sample,' which is generated. These errors must be estimated and then a 'policy' decision must be made, that is, should more CPU time be used to reduce the statistical errors or should the CPU time available.

1.3.1.2.3   WHAT STRATEGY SHOULD FOLLOWED IN APPROACHING A PROBLEM?

Most new simulations face hidden pitfalls and difficulties, which may not be apparent in early phases of the work. It is therefore often advisable to begin with a relatively simple program and use relatively small system sizes and modest running times. Sometimes there are special values of parameters for which the answers are already known (either from analytic solutions or from previous, high quality simulations) and these cases can be used to test a new simulation program. By proceeding in this manner one is able to uncover which are the parameter ranges of interest and what unexpected difficulties are present. It is then possible to refine the program and then to increase running times. Thus bothCPU time and human time can be used most.

1.3.1.2.4   HOW DO SIMULATIONS RELATE TO THEORY AND EXPERIMENTAL WORKS?

In many cases theoretical treatments are available for models for which there is no perfect physical realization (at least at the present time). In this situation the only possible test for an approximate theoretical solution is to compare with 'data' generated from a computer simulation. As an example we wish to mention recent activity in growth models, such as diffusion-limited aggregation, for which a very large body of simulation results already exists but for which extensive experimental information is just now becoming available. It is not an exaggeration to say that interest in this field was created by simulations. Even more dramatic examples are those of reactor meltdown or large-scale nuclear war: although we want to know what the results of such events would be we do not want to carry out experiments! There are also real physical systems, which are sufficiently complex that they are not presently amenable to theoretical treatment. An example is the problem of understanding the specific behavior of a system with many competing interactions and which is undergoing a phase transition. A model Hamiltonian, which is believed to contain all the essential features of the physics may be proposed, and its properties may then be determined from simulations. If the simulation (which now plays the role of theory) disagrees with experiment, then a new Hamiltonian must be sought. An important advantage of the simulations is that different physical effects, which are simultaneously present in real systems may be isolated and through separate consideration by simulation may provide a much better understanding.

1.3.1.2.5   THE ART OF RANDOM NUMBER GENERATION, BACKGROUND

Monte Carlo methods are heavily dependent on the fast, efficient production of streams of random numbers. Since physical processes, such as white noise generation from electrical circuits, generally introduce new numbers much too slowly to be effective with today's digital computers, random number sequences are produced directly on the computer using software (the use of tables of random numbers is also impractical because of the huge number of random numbers now needed for most simulations and the slow access time to secondary storage media). Since such algorithms are actually deterministic, the random number sequences, which are thus produced are only 'pseudorandom' and do indeed have limitations, which need to be understood. Thus, in the Appendix of this review, when we refer to Generation 'Random numbers' programs 1 and 2, Some necessary background 'random numbers' it must be understood that we are really speaking of 'pseudorandom'numbers (Appendix A).

These deterministic features are not always negative. For example, for testing a program it is often useful to compare the results with a previous run made using exactly the same random numbers. The explosive growth in the use of Monte Carlo simulations in diverse areas of physics has prompted extensive investigation of new methods and of the reliability of both old and new techniques. Monte Carlo simulations are subject to both statistical and systematic errors from multiple sources, some of which are well understood. It has long been known that poor quality random number generation can lead to systematic errors in Monte Carlo simulation; in fact, early problems with popular generators led to the development of improved methods for producing pseudorandom numbers. As we shall show in the following discussion both the testing as well as the generation of random numbers remain important problems that have not been fully solved. In general, the random number sequences, which are needed, should be uniform, uncorrelated, and of extremely long period, that is, do not repeat over quite long intervals. In the following subsections we shall discuss several different kinds of generators. The reason for this is that it is now clear that for optimum performance and accuracy, the random number generator needs to be matched to the algorithm and computer. Indeed, the resolution of Monten Carlo studies has now advanced to the point where no generator can be considered to be completely 'safe' for use with a new simulation algorithm on a new problem. The practitioner is now faced anew with the challenge of testing the random number generator for each high-resolution application, and we shall review some of the 'tests' later in this section. The generators, which are discussed in the next subsections produce a sequence of random integers. One important topic, which we shall not consider here is the question of the implementation of random number generators on massively parallel computers. In such cases one must be certain that the random number sequences on all processors are distinct and uncorrelated. As the number of processors available to single users increases, this question must surely be addressed, but we feel that at the present time this is a rather specialized topic and we shall not consider it further. This method for generation of random walk numbers including: Congruential method, Mixed congruential methods, Shift register algorithms, Lagged Fibonacci generators, Tests for quality, non-uniform distributions.

1.3.2 MOLECULAR DYNAMICS SIMULATION AND ITS AIMS

Simulations as a bridge between microscopic and macroscopic, theory and experiment. We carry out computer simulations in the hope of understanding the properties of assemblies of molecules in terms of their structure and the microscopic interactions between them. This serves as a complement to conventional experiments, enabling us to learn something new, something that cannot be found out in other ways. In this review we shall concentrate on MD. The obvious advantage of MD over MC is that it gives a route to dynamical properties of the system: transport coefficients, time-dependent responses to perturbations, rheological properties and spectra. Computer simulations act as a bridge between microscopic length and timescales and the macroscopic world of the laboratory: we provide a guess at the interactions between molecules, and obtain 'exact' predictions of bulk properties. The predictions are 'exact' in the sense that they can be made as accurate as we like, subject to the limitations imposed by our computer budget. At the same time, the hidden detail behind bulk measurements can be revealed. An example is the link between the diffusion coefficient and velocity autocorrelation function (the former easy to measure experimentally, the latter much harder). Simulations act as a bridge in another sense: between theory and experiment. We may test a theory by conducting a simulation using the same model. We may test the model by comparing with experimental results. We may also carry out simulations on the computer that are difficult or impossible in the laboratory (for example, working at extremes of temperature or pressure). Ultimately we may want to make direct comparisons with experimental measurements made on species materials, in which case a good model of molecular interactions is essential. The aim of so-called ab initio molecular dynamics is to reduce the amount of tting and guesswork in this process to a minimum. On the other hand, we may be interested in phenomena of a rather generic nature, or we may simply want to discriminate between good and bad theories. When it comes to aims of this kind, it is not necessary to have a perfectly realistic molecular model; one that contains the essential physics may be quite suitable [122, 132].

1.3.2.1 MILESTONES IN MD SIMULATIONS OF SWNT FORMATION

As we revisit the milestones in MD simulations of SWNT nucleation and growth, the degree to which each investigation satisfies each of these four hypothetical "tasks" will be emphasized [129].

1.3.2.1.1 REBO-BASED MD SIMULATIONS

The Ters off potential and its application to hydrocarbon systems, called the REBO potential, describes bond breaking/formation using a distance-dependent many-body bond order term and correctly dissociating attractive/repulsive diatomic potentials. This approach is nevertheless based on a local molecular mechanics approach, and therefore lacks explicit descriptions of quantum phenomena and Coulomb/vander Waals interactions. Hence, changes in the electronic structure of a dynamic, molecular system are not described. In particular, neither the evolution of conjugation and aromatic stabilization of carbons (important for sp^2-hybridized carbon), nor charge transfer effects and the near degeneracy of metal d-orbitals (important for the catalyst) can be captured by MD simulations based on the Ters off and REBO potentials. Besides these grave decencies, the standard REBO potential has several other well-known problems. For example, in modeling gas-phase carbon densities it overestimates the sp^3 fraction compared to corresponding DFT simulations, while it underestimates the sp fraction which is of paramount importance with respect to fullerene cage self-assembly [129].

1.3.2.1.2 CPMD SIMULATIONS

CPMD simulations have been employed by two research groups for simulating the growth process of metal-catalyzed SWNT growth. Remarkably, the study by researchers, was performed even before the first REBO-based MD simulations by the Maruyama group. As stated above, CPMD is a more accurate MD approach compared to REBO based MD, since it includes changes in the electronic structure throughout the dynamics. However, the reported CPMD studies were completely inadequate because of their short simulation time (due to high computational cost) and their unrealistic initial model geometries. Based on researchers' limited simulation results, researchers suggested that the segregation of carbon linear chains and atomic rings on the surface of a liquid-like cobalt–carbide particle represents the first stage of the cap nucleation process. In the many papers, researchers, also simulated the incorporation of ve carbon atoms into a preassembled half-fullerene cap attached to a cobalt metal surface for 15 ps. CPMD is an ab initio molecular dynamics method which is the combination of first principles electronic structure methods with MD based on Newton's equations of motion. Grand-state electronic structures were described according to DFT within plane-wave pseudo potential framework. The use of electronic structure methods to calculate the interaction potential between atoms overcomes the main shortcomings of the otherwise highly successful pair potential approach. There have been plenty of excellent reference books on MD and DFT, and some simulation tricks can be found. Here, some more details about CPMD, which are different from the traditional classical MD and DFT will be discussed. In CPMD, considering the parameters {wi}, {RI}, {av} in energy function are supposed to be time-dependent, the Lagrangean:

$$L = \sum_i \frac{1}{2}\mu \int_{\dot{U}} d^3 r |\psi_i|^2 + \sum_I \frac{1}{2}M_I R_I^2 + \sum_v \frac{1}{2}\mu_v \alpha_v^2 - E\left[\{\psi_i\},\{R_I\},\{\alpha_v\}\right] \qquad (96)$$

$$L = \sum_i \frac{1}{2}\mu \int_{\dot{U}} d^3 r |\psi_i|^2 + \sum_I \frac{1}{2}M_I R_I^2 + \sum_v \frac{1}{2}\mu_v \alpha_v^2 - E\left[\{\psi_i\},\{R_I\},\{\alpha_v\}\right] \qquad (97)$$

was introduced, where the w_i are subject to the holonomic constraints:

$$\sum_i \int_{\dot{U}} d^3 r \psi_i^* (r,t)\psi_j (r,t) = \delta_{ij} \qquad (98)$$

In Eqs. (96) and (97), w_i are orbitals for electrons, RI indicate the nuclear coordinates and a_v are all the possible external constraints imposed on the system, $w^*(r)$ is the complex conjugate of wave function $w(r)$, h is the reduced Planck constant, m is the mass of electron and $n(r) = Ri |wi(r)|^2$ is the electron density; the dot indicates time derivative, M_I are the physical ionic masses. Then, the equations of motion can be written as:

$$\mu \ddot{\psi}_i (r,t) = -\frac{\delta E}{\delta \psi_i^* (r,t)} + \sum_i \Lambda_{ik} \psi_K (r,t) \qquad (99)$$

$$M_I \ddot{R}_I = -\nabla_{RI} E \qquad (100)$$

$$\mu_v \ddot{\alpha}_v = -\left(\frac{\partial E}{\partial \alpha_v}\right) \qquad (101)$$

That is Lagrange multipliers introduced in order to satisfy the constraints in Eq. (99). Then the equation of kinetic energy:

$$K = \sum_i \frac{1}{2} \mu \int_{\dot{U}} d^3 r \left| \dot{\psi}_i \right|^2 + \sum_I \frac{1}{2} M_I \dot{R}_I^2 + \sum_v \frac{1}{2} \mu_v \dot{\alpha}_v^2 \tag{102}$$

is obtained. Based on the technique mentioned, CPMD extends MD beyond the usual pair-potential approximation. In addition, it also extends the application of DFT to much larger systems [129, 141].

1.3.2.1.3   HYBRID METHODS (MD + MC)

For some complex systems Monte Carlo simulations have very low acceptance rates except for very small trial moves and hence become quite inefficient. Molecular dynamics simulations may not allow the system to develop sufficiently in time to be useful, however, molecular dynamics methods may actually improve a Monte Carlo investigation of the system. A trial move is produced by allowing the molecular dynamics equations of motion to progress the system through a rather large time step. Although such a development may no longer be accurate as a molecular dynamics step, it will produce a Monte Carlo trial move, which will have a much higher chance of success than a randomly chosen trial move. In the actual implementation of this method some testing is generally advisable to determine an effective value of the time step.

1.3.2.1.4   AB INITIO MOLECULAR DYNAMICS

No discussion of molecular dynamics would be complete without at least a brief mention of the approach pioneered by many researchers, which combines electronic structure methods with classical molecular dynamics. In this hybrid scheme a fictitious dynamical system is simulated in which the potential energy is a functional of both electronic and ionic degrees of freedom. This energy functional is minimized with respect to the electronic degrees of freedom to obtain the potential energy surface to be used in solving for the trajectories of the nuclei. This approach has proven to be quite fruitful with the use of density functional theory for the solution of the electronic structure part of the problem and appropriately chosen pseudo potentials. These equations of motion in relation with molecular dynamics simulation, can then be solved by the usual numerical methods, for example, the Verlet algorithm, and constant temperature simulations can be performed by introducing thermostats or velocity rescaling. This ab initio method is efficient in exploring complicated energy landscapes in whichboth the ionic positions and electronic structures are determined simultaneously.

1.3.2.1.5   DFTB-BASED MD SIMULATIONS

Neither REBO-based nor first principle-based MD simulations have succeeded in modeling the nucleation and sustained growth of a clean hexagononly SWNT from scratch as routinely achieved experimentally. This is despite the use of such tricks as injecting carbon atoms into the middle of the metal catalyst particle. It seems that one should "bite the bullet" and deal somehow with the complicated electronic structure of the evolving sp^2-network and transition metal clusters at lower computational cost. The major obstacle for such a "cheap" CPMD approach is the limited availability of metal parameters in conventional semi empirical quantum chemical methods. Most notable in this respect are the MNDO/d or PM6 codes. However, the metal parameters are typically designed for single metal atom systems and are not suitable for the treatment of metal nanoparticles. The DFTB/MD approach, which is a quantum mechanics/molecular dynamics (QM/MD) technique based on the DFTB electronic structure method, can play a role to in a gap between classical and the principle MD simulations. The DFTB method is approximately two orders of magnitude faster than first principles DFT and therefore enables longer simulations and provides more adequate model systems for non-equilibrium dynamics of nanosized clusters with quantum mechanical treatment of electrons. In addition, metal–carbon parameters for DFTB have recently been developed by the Morokuma group. A nite electronic temperature approach ensures the applicability of the DFTB/MD method for nanometer size metal particles with high electronic densities of states around the Fermi level, as it allows the occupancy of each molecular orbital to change smoothly from 2 to 0 depending on its orbital energy. This approach effectively incorporates the open shell nature of the system due to near-degeneracy of iron d-orbitals, as well as carbon dangling bonds. We have found that in the absence of electronic temperature the iron cluster is much less reactive. This is considered to be an artifact of the simulation of a near-closed shell electronic wave function in that case Initially, we employed a (5,5) armchair SWNT fragment attached to an Fe_{38} cluster as a model system. The open end of the SWNT seed was terminated by hydrogen atoms, whereas the other end was bound to surface iron atoms of the Fe_{38}cluster [129, 144].

1.3.2.1.6   SIMULATION OF ADSORBED MONOLAYERS IN SMOOTH SUBSTRATES

The study of two-dimensional systems of adsorbed atoms has attracted great attention because of the entire question of the nature of two-dimensional melting. In the absence of a periodic substrate potential, the system is free to form an ordered structure determined solely by the inter particle interactions. As the temperature is raised this planar 'solid' is expected to melt, but the nature of the transition is a matter of debate 6.4 adsorbed monolayers 205 Fast multipole method:

(1) Divide the system into sets of successively smaller Sub- Simulation Cells.
(2) Shift the origin of the multipole expansion and calculate the multipole moments at all subcell levels starting from the lowest level.
(3) Shift the origin of the local expansion and calculate the local expansion coefficients starting from the highest level.
(4) Evaluate the potential and fields for each particle using local expansion coefficients for the smallest subcell containing the particle.
(5) Add the contributions from other charges in the same cell and near neighbor cells by direct summation.

1.3.2.1.7 PERIODIC SUBSTRATE POTENTIALS

Extensive experimental data now exist for adsorbed monolayers on various crystalline substrates and there have been a number of different attempts made to carry out simulations, which would describe the experimental observations. These fall into two general categories: lattice gas models, and off lattice models with continuous, position dependent potentials. For certain general features of the phase diagrams lattice gas models offer a simple and exceedingly efficient simulations capability. This approach can describe the general features of order or disorder transitions involving commensurate phases. An extension of the lattice gas description for the ordering of hydrogen on palladium in the C(2 2) structure has recently been proposed by giving the ad atoms translational degrees of freedom within a lattice cell (Presber et al., 1998).The situation is complicated if one wishes to consider orientational transitions involving adsorbed molecules since continuous degrees of freedom must be used to describe the angular variables. Both quadrupolar and octu polar systems have been simulated. For a more complete description of the properties of adsorbed monolayers it is necessary to allow continuous movement of particles in a periodic potential produced by the underlying substrate. One simplification, which is often used is to constrain the system to lie in a two-dimensional plane so that the height of the ad atoms above the substrate is fixed. The problem is still difficult computationally since there may be strong competition between ordering due to the adatom or adatom interaction and the substrate potential and incommensurate phases may result. Molecular dynamics has been used extensively for this class of problems but there have been Monte Carlo studies as well. One of the 'classic' adsorbed monolayer systems is Kr on graphite. The substrate has hexagonal symmetry with a lattice constant of 2.46A whereas the lattice constant of a compressed two-dimensional krypton solid is 1.9A. The C(1 1) structure is thus highly unfavorable and instead we find occupation of next-nearest neighbor graphite hexagons leading to a commensurate structure with lattice constant 4.26 A. This means, however, that the krypton structure must expand relative to an isolated two-dimensional solid. Thus, there is competition between the length scales set by the Kr±Kr and Kr±graphite interactions. An important question was thus whether or not this competition could lead to an in commensurate phase at low temperatures. This is a situation in which boundary conditions again become an important consideration. If periodic boundary conditions are imposed, they will naturally tend to produce a structure, which is periodic with respect to the size of the simulation cell. In this case a more profitable strategy is to use free edges to provide the system with more freedom. The negative aspect of this choice is that finite size effects become even more pronounced. This question has been studied using a Hamiltonian and some mentioned equations.

1.3.2.1.8   COMPARISON OF SOME SIMULATION METHODS

As noted from the above works, the large numbers of numerical studies on CNT dynamics were based on MD simulations. In the classical MD simulations, all degrees of freedom due to the electrons are ignored, as well as quantum effects. So a more accurate description is needed, i.e., potentials between electron–electron, electron–ion as well as ion–ion interactions should be considered. In this respect, the ab initio density functional theory (DFT) and Car–Parrinello molecular dynamics (CPMD) were used in the study. DFT calculation, which is known as time consuming and considering more details about the electrons and the ions, can give more interactions and corrections about electron–electron, electron– ion and ion–ion, which are disregarded in classical MD. Considered these corrections, the simulated Young's moduli of CNTs would be more credible. But it can only simulate the static state. CPMD, which combines the MD simulation with quantum mechanics, treats the electronic degrees of freedom in the framework of DFT. It makes a balance between efficiency and precision, while considering the dynamic effect. However, the dynamic processes, performing in the CPMD still restricted to small system, which is a main disadvantage of CPMD. Here compendiously compared the threedifferent methods in Table 1.6 [129]. It is apparent from non-hexagonal SWNT with a (n,m) chirality, cannot be achieved using REBO-based MD simulations. The effect of the increase of the conjugated electronic structure with increased stabilization is beyond the capability of simple many-body potentials. Further, the second chief objective of the metal cluster, namely keeping the carbon structure open for continued sidewall growth by incoming new carbon feedstock cannot be achieved without "cheating": that is, a supply of carbon from the outside of the metal cluster always results in complete encapsulation and "death of the catalyst".

TABLE 1.6   Brief Comparison of Three Different Simulation Methods

	Merits	Drawbacks
DFT	High accuracy; more details (electronic states, charge distribution, molecule orbits)	Limited to static states of small systems; slow and expensive
MD	Available for large systems; fast and cheap	Disable in chemical reaction (bond breaking/forming); empirical potentials are used whichleads to low accuracy
CPMD	Combined MD with DFT; a balance between the time consuming andprecision	Limited to dynamic process of small systems

1.3.2.2 MOLECULAR INTERACTIONS

Molecular dynamics simulation consists of the numerical, step-by-step, solution of the classical equations of motion, which for a simple atomic system may be written:

$$m_i \ddot{r}_i = f_i, f_i = -\frac{\partial}{\partial r_i} U$$

(103)

For this purpose we need to be able to calculate the forces fi acting on the atoms, and these are usually derived from a potential energy $U(r_N)$, where $r_N = (r_1; r_2; : : : r_N)$ represents the complete set of 3N atomic coordinates. In this section, we focus on this function $U(r_N)$, restricting ourselves to an atomic description for simplicity. In simulating soft condensed matter systems, we sometimes wish to consider non-spherical rigid units, which have rotational degrees of freedom.

1.3.2.3 NON-BONDED INTERACTIONS

The part of the potential energy U non-bonded representing non-bonded interactions between atoms is traditionally split into 1-body, 2-body, 3-body, ... terms:

$$U_{non-bonded}\left(r^N\right) = \sum_i u\left(r_i\right) + \sum_i \sum_{j>i} v\left(r_i, r_j\right) + \cdots$$

(104)

The u(r) term represents an externally applied potential eldor the effects of the container walls; it is usually dropped for fully periodic simulations of bulk systems. Also, it is usual to concentrate on the pair potential $v(r_i; r_j) = v(r_{ij})$ and neglect three-body (and higher order) interactions. There is an extensive literature on the way these potentials are determined experimentally, or modeled theoretically. In some simulations of complex fluids, it is sufficient to use the simplest models that faithfully represent the essential physics. In this chapter we shall concentrate on continuous, differentiable pair-potentials (although discontinuous potentials such as hard spheresand spheroids have also played a role). The Lennard-Jones potential is the most commonly used form:

$$v^{l,J}(r) = 4\varepsilon \left[\left(\frac{\sigma}{r} \right)^{12} - \left(\frac{\sigma}{r} \right)^{6} \right] \tag{105}$$

with two parameters—the diameter, and the well depth. This potential was used, for instance, in the earliest studies of the properties of liquid argon. For applications in which attractive interactions are of less concern than the excluded volume effects, which dictate molecular packing, the potential may be truncatedat the position of its minimum, and shifted upwards to give what is usually termed the model. If electrostatic charges are present, we add the appropriate Coulomb potentials.

$$v^{Coulomb}(r) = \frac{Q_1 Q_2}{4\pi \in_0 r} \tag{106}$$

where Q_1, Q_2 are the charges and is the permittivity of free space. The correct handling of long-range forces in a simulation is an essential aspect of polyelectrolyte simulations, which will be the subject of the later chapter of Holm [145].

1.3.2.4   BONDING POTENTIALS

For molecular systems, we simply build the molecules out of site-site potentials of the form of Eq. (69) or similar. Typically, a single-molecule quantum-chemical calculation may be used to estimate the electron density throughout the molecule, whichmay then be modeled by a distribution of partial charges via Eq. (70), or more accurately by a distribution of electrostatic multipoles. For molecules we must also consider the intramolecular bonding interactions. The simplest molecular model will include terms of the following kind:

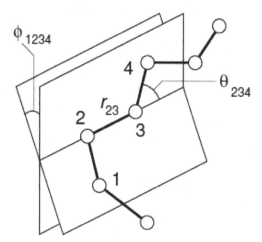

FIGURE 1.6 Geometry of a simple chain molecule, illustrating the definition of interatomic distance r_{23}, bend angle, and torsion angle .

$$U_{intramolecular} = \frac{1}{2}\sum_{bonds} k_{ij}^r \left(r_{ij} - r_{eq} \right)^2 \tag{107}$$

$$+\frac{1}{2}\sum_{\substack{bonds \\ angles}} k_{ijk}^0 \left(0_{ijk} - 0_{eq} \right)^2 \tag{108}$$

$$+\frac{1}{2}\sum_{\substack{torsion \\ angles}}\sum_m k_{ijkl}^{\phi,m} \left(1 + \cos(m\phi_{ijkl} - \gamma_m \right)^2 \tag{109}$$

The geometry is illustrated in Fig. 1.6. The "bonds" will typically involve the separation between adjacent pairs of atoms in a molecular framework, and we assume inEq. (107), a harmonic form with specified equilibrium separation, although this is not the only possibility. The "bend angles"are between successive bond vectors such as and , and therefore involve three atom coordinates:

$$\cos\theta_{ijk} = \hat{r}_{ij}.\hat{r}_{jk} = \left(r_{ij}.r_{ij} \right)^{-1/2} \left(r_{jk}.r_{jk} \right)^{-1/2} \left(r_{jk}.r_{jk} \right) \tag{110}$$

where Usually this bending term is taken to be quadratic in the angular displacement from the equilibrium value, as in Eq. (110), although periodic functions are also used. The "torsion angles"ijkl are defined in terms of three connected bonds, hence four atomic coordinates:

$$\cos\phi_{ijkl} = -\hat{n}_{ijk}.\hat{n}_{jkl}, where\, n_{ijk} = r_{ij} \times r_{jk}, n_{jkl} = r_{jk} \times r_{kl} \tag{111}$$

and , the unit normal to the plane dened by each pair of bonds. Usually the torsional potential involves an expansion in periodic functions of order Eq. (109).

A simulation package force-eld will specify the precise form this equation. Molecular mechanics force fields, aimed at accurately predicting structures and properties, will include many cross-terms (e.g., stretch-bend). Quantum mechanical calculations may give a guide to the "best" molecular force-eld; also comparison of simulation results with thermophysical properties and vibration frequencies is invaluable in force-eld development and tenement. A separate family of force elds, such as AMBER, CHARMM and OPLS are geared more to larger molecules (proteins, polymers) in condensed phases; their functional form is simpler, closer to that of this equation and their parameters are typically determined by quantum chemical calculations combined with thermophysical and phase coexistence data. This eld is too broad to be reviewed here; several molecular modeling texts (albeit targeted at biological applications) should be consulted by the interested reader. The modeling of long chain molecules will be of particular interest to us, especially as an illustration of the scope for progressively simplifying and "coarse-graining" the potential model. Various explicit-atom potentials have been devised for the n-alkanes. More approximate potentials have also been constructed in which the CH_2 and CH_3 units are represented by single "united atoms." These potentials are typically less accurate and less transferable than the explicit-atom potentials, but significantly less expensive; comparisons have been made between the two approaches. For more complicated molecules this approach may need to be modified. In the liquid crystal eld, for instance, a compromise has been suggested: use the united-atom approach for hydrocarbon chains, but model phenyl ring hydrogen's explicitly. In polymer simulations, there is frequently a need to economize further and coarse grain the interactions more dramatically: significant progress has been made in recent years in approaching this problem systematically. Finally, the most fundamental properties, such as the entanglement length in a polymer melt, may be investigated using a simple chain of pseudoatoms or beads (modeled using the WCA potential and each representing several monomers), joined by an attractive nitely extensible non-linear elastic (FENE) potential:

$$v^{\text{FENE}}(r) = \begin{cases} -\dfrac{1}{2}kR_0^2 \ln \ln\left(1-\left(\dfrac{r}{R_0}\right)^2\right) & r < R_0 \\ \infty & r < R_0 \end{cases} \qquad (112)$$

The key feature of this potential is that it cannot be extended beyond , ensuring (for suitable choices of the parameters k and R_0 that polymer chains cannot move through one another [132].

1.3.2.5   PERIODIC BOUNDARY CONDITIONS

Small sample size means that, unless surface effects are of particular interest, periodic boundary conditions need to be used. Consider 1000 atoms arranged in a cube. Nearly half the atoms are on the outer faces, and these will have a large effect on the measured properties. Even for $10^6 = 100^3$ atoms, the surface atoms amount to 6% of the total, which is still non-trivial. Surrounding the cube with replicas of itself takes care of this problem. Provided the potential range is not too long, we can adopt the minimum image convention that each atom interacts with the nearest atom or image in the periodic array. In the course of the simulation, if an atom leaves the basic simulation box, attention can be switched to the incoming image. Of course, it is important to bear in mind the imposed artificial periodicity when considering properties, which are influenced by long-range correlations. Special attention must be paid to the case where the potential range is not short: For example, for charged and dipolar systems.

1.3.2.6   NEIGHBOR LISTS

Computing the non-bonded contribution to the inter atomic forces in an MD simulation involves, in principle, a large number of pair wise calculations: we consider each atom and loop over all other atoms j to calculate the minimum image separations . Some economies result from the use of lists of nearby pairs of atoms. Verlet suggested such a technique for improving the speed of a program. The potential cutoff sphere, of radius, around a particular atom is surrounded by a 'skin,' to give a larger sphere of radius. At the first step in a simulation, a list is constructed of all the neighbors of each atom, for which the pair separation is within . Over the next few MD time steps, only pairs appearing in the list are checked in the force routine.

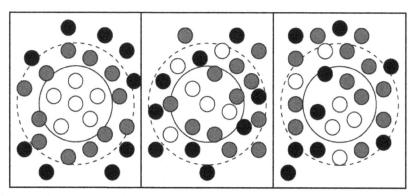

FIGURE 1.7   The Verlet list on its construction, later, and too late. The potential cutoff range (solid circle), and the list range (dashed circle), are indicated. The list must be reconstructed before particles originally outside the list range (black) have penetrated the potential cutoff sphere.

Time-to-time the list is reconstructed: it is important to do this before any unlisted pairs have crossed the safety zone and come within interaction range. It is possible to trigger the list reconstruction automatically, if a record is kept of the distance traveled by each atom since the last update. The choice of list cutoff distance is a compromise: larger lists will need to be reconstructed less frequently, but will not give as much of a saving on CPU time as smaller lists. This choice can easily be made by experimentation. For larger systems (or so, depending on the potential range) another technique becomes preferable. The cubic simulation box (extension to non-cubic cases is possible) is divided into a regular lattice of cells. The first part of the method involves sorting all the atoms into their appropriate cells. This sorting is rapid, andmay be performed every step. Then, within the force routine, pointers are used to scan through the contents of cells, and calculate pair forces. This approach is very efficient for large systems with short-range forces. A certain amount of unnecessary work is done because the search region is cubic, not (as for the Verlet list) spherical. Molecular dynamics evolves a nite-sized molecular configuration forward in time, in a step-by-step fashion. There are limits on the typical time scales and length scales that can be investigated and the consequences must be considered in analyzing the results. Simulation runs are typically short: typicallyMD steps, corresponding to perhaps a few nanoseconds of real time, and in special cases extending to the microsecond regime. This means that we need to test whether or not a simulation has reached equilibrium before we can trust the averages calculated in it. Moreover, there is a clear need to subject the simulation averages to a statistical analysis, to make a realistic estimate of the errors. How long should we run? This depends on the system and the physical properties of interest as the variables and become uncorrelated; this decay occurs over a characteristic time . Formally we may define a correlation time:

$$\tau_a = \int\limits_0^\infty \frac{dt a(0) a(t)}{a^2} \qquad (113)$$

It is almost essential for simulation box sizes to be large and for simulation run lengthsto be large compared with , for all properties of interest Only then can we guarantee that reliablysampled statistical properties are obtained. Roughly speaking, the statistical error in a property calculated as an average over a simulation run of length is proportional to the time average is essentially a sum of independent quantities, each an average over time. Near critical points, special care must be taken, in that these inequalities will almost certainly not be satisfied, and indeed one may see the onset of non-exponential decay of the correlation functions. In these circumstances a quantitative investigation of nite size effects and correlation times, with some consideration of the appropriate scaling laws, must be undertaken. Phase diagrams of soft-matter systems often include continuous phase transitions, or weakly first-order transitions exhibiting significant pretransitional actuations. One of the most encouraging developments of recent years has been the establishment of reliable and systematic methods of studying critical phenomena by simulation, although typically the Monte Carlo method is more useful for this type of study.

1.3.2.7 PREDICTING GAS DIFFUSIVITY USING MOLECULAR DYNAMICS SIMULATIONS IN MOFS

1.3.2.7.1 EQUILIBRIUM MOLECULAR DYNAMICS SIMULATIONS IN THIS CASE STUDY

Over approximately the last decade, metal organic framework (MOF) materials have attracted a great deal of attention as a new addition to the classes of nanoporous materials. MOFs, also known as porous coordination polymers (PCPs) or porous coordination networks (PCNs), are hybrid materials composed of single metal ions or polynuclear metal clusters linked by organic ligands through strong coordination bonds. Due to these strong coordination bonds, MOFs are crystallographically well-defined structures that can kept their permanent porosity and crystal structure after the removal of the guest species used during synthesis. In the literature, MD simulations have been used to predict three different types of gas diffusivities in MOFs. These are transport diffusivity, corrected diffusivity and self-diffusivity. MOFs have become attractive alternatives to traditional nanoporous materials specifically in gas storage and gas separation since their synthesis can be readily adapted to control pore connectivity, structure and dimension by varying the linkers, ligands and metals in the material. The enormous number of different possible MOFs indicates that purely experimental means for designing optimal MOFs for targeted applications is inefficient at best. Efforts to predict the performance of MOFs using molecular modeling play an important role in selecting materials for specific applications. In many applications that are envisioned for MOFs, diffusion behavior of gases is of paramount importance. Applications such as catalysis, membranes and sensors cannot be evaluated for MOFs without information on gas diffusion rates. Most of the information on gas diffusion in MOFs has been provided by molecular dynamics (MD) studies. The objective of this chapter is to review the recent advances in MD simulations of gas diffusion in MOFs. In Sections 1.3.2.7.2 and 1.3.2.7.3 the MD models used for gas molecules and MOFs will be introduced. Studies which computed single component and mixture gas diffusivities in MOFs will be reviewed in Section 1.3. The discussion of comparing results of MD simulations with the experimental measurements and with the predictions of theoretical correlations will be given in Section 1.2.2.11, respectively. Finally, opportunities and challenges in using MD simulations for examining gas diffusion in MOFs will be summarized in Section Gas diffusion is an observable consequence of the motion of atoms and molecules as a response to external force such as temperature, pressure or concentration change. Molecular dynamics (MD) is a natural method to simulate the motion and dynamics of atoms and molecules. The main concept in an MD simulation is to generate successive configurations of a system by integrating Newton's law of motion. Using MD simulations, various diffusion coefficients can be measured from the trajectories showing how the positions and velocities of the particles vary with time in the system. Several different types of gas diffusion coefficients and the methods to measure them will be addressed in the next section in details. In accessing the gas diffusion in nanoporous materials, equilibrium MD simulations, which model the behavior of the system in equilibrium, have been very widely used. In equilibrium MD simulations, first a short grand canonical Monte Carlo (GCMC) simulation is applied to generate the initial configurations of the atoms in the nanopores. Initial velocities are generally randomly assigned to each particle (atom) based on Maxwell-Boltzmann velocity distribution. An initial NVT-MD (NVT: constant number of molecules, constant volume, constant temperature) simulation is performed to equilibrate the system. After the equilibration, Newton's equation is integrated and the positions of each particle in the system are recorded at a prespecified rate. Nosé-Ho over thermostat is very widely applied to keep the desired temperature and the integration of the system dynamics is based on the explicit N-V-T chain integrator by keeping temperature constant, Newton's equations are integrated in a canonical ensemble (NVT) instead of a microcanonical ensemble. To describe the dynamics of rigid-linear molecules such as carbon dioxide the MD algorithm is widely used. The so-called order N algorithm is implemented to calculate the diffusivities from the saved trajectories. In order to perform classical MD simulations to measure gas diffusion in MOFs' pores, force fields defining interactions between gas molecules-gas molecules and gas molecules-MOF's atoms are required. Once these force fields are specified, dynamical properties of the gases in the simulated material can be probed. These force fields will be studied in two parts: models for gas molecules (adsorbates) and models for MOFs (adsorbents).

1.3.2.7.2   MODELS FOR GASES

Diffusion of hydrogen, methane, argon, carbon dioxide and nitrogen are very widely studied in MOFs. For H_2, three different types of fluid-fluid potential models have been used. In most of the MD simulations, spherical 12-6 Lennard-Jones (LJ) model has been used for H_2. The Buch potential is known to reproduce the experimental bulk equation of state accurately for H_2. Two-site LJ models have also been used in the literature. The potential model of many researchers has been used to account for the quadrupole moment of H_2 molecules. This potential consists of a LJ core placed at the center of mass of the molecule and point charges at the position of the two protons and the center of mass.

1.3.2.7.3   MODELS FOR MOFS

When the first MD simulations were performed to examine gas diffusion in MOFs at the beginning of 2004, there was no experimental data to validate the accuracy of

MD studies. However, in general whenever experimental equilibrium properties such as adsorption isotherms have been reproduced by the molecular simulations, it has been observed that dynamic simulations based on the same interatomic potentials are also reliable. Therefore, many MD studies examining gas diffusion in MOFs first showed the good agreement between experiments and simulations for gas adsorption isotherms and then used the same potential models for gas diffusion simulations. Here, it is useful to highlight that considering a wide gas loading range when comparing simulation results with experimental data is crucial. It is unreasonable tocompare outcome of simulations with the experimental measurements over a very narrow range of loading and assumes that good (or poor) agreement with experiment will continue to high loadings. The MD simulations have used general-purpose force fields. The MD simulations have used general-purpose force fields such as the universal force field (UFF), DREIDING force field(and optimized potential for liquid simulations all-atom (OPLS-AA) force field for representing the interactions between MOF atoms and adsorbates. There are studies where the parameters of the force fields are refined to match the predictions of simulations with the experimental measurements (in most cases experimental adsorption isotherm data existwhereas experimental diffusion data do not exist) or using first principles calculations. Of course, one must be careful in refining force field parameters to match the results of simulations with the experimental data since the accuracy of the experiments are significantly affected by the defects of as synthesized MOFs or trapped residual solvent molecules present in the samples. Most MD simulations performed to date have assumed rigid MOF structures, which mean the framework atoms are fixed at their crystallographic positions. Generally, the crystallographic data for MOFs are obtained from X-ray diffraction experiments. In rigid framework simulations, only the non-bonding parameters, describing the pair wise interactions between the adsorbate and the adsorbent atoms of the particular force field, were used. It can be anticipated that the assumption of a rigid framework brings a huge computational efficiency yet the inclusion of the lattice motion and deformation is crucial for an accurate description of diffusion of large gas molecules since they fit tightly in the MOF pores, forcing the MOF to deform in order to allow migration from pore to pore. The literature summary presented so far indicates that the number of MD simulation studies with flexible MOFs and flexible force fields is very limited. More research will sure be helpful to understand the importance of lattice dynamics on diffusivity of gas molecules in MOFs. Studies to date indicated that the lattice dynamics are specifically important in computing diffusivity of large gas molecules (such as benzene) in MOFs having relatively narrow pores. Studies on flexible force fields also suggested that a force field developed for a specific MOF can be adapted to similar MOF structures (as in the case of IRMOFs) with slight modifications for doing comparative studies to provide a comprehensive understanding of gas diffusion in flexible MOFs.

1.3.2.8 DUAL-CONTROL-VOLUME GRAND CANONICAL MOLECULAR DYNAMIC (DCV-GCMD) METHOD FOR PORE SIZE EFFECTS ANALYSIS ON DIFFUSION IN SWCNTS

Dual-control-volume grand canonical molecular dynamics simulations were used to study the diffusion mechanisms of counter diffusing CH/CF_4 mixtures in cylindrical model Pores. It was found that in the Pores two different diffusion mechanisms occur independently of each other. Therefore, the pores were divided for purposes of analysis into two regions, a wall region close to the pore wall where most of the fluid molecules are located and an inner region where fewer molecules are Present but from where the main contribution to the flux comes. The dependence of the transport diffusion coefficients on pore radius and temperature were analyzed separately for these two regions. The varying contributions from fluid-fluid and fluid-wall collisions to the diffusion mechanism could be demonstrated. Whereas in the wall region surface diffusion takes Place, in the inner region diffusion occurs in the transition regime between molecular and Knudsen diffusion. Depending on different factors such as the pore size, or the temperature different diffusion mechanisms apply. There exist several simulation studies where these different diffusion mechanisms are investigated by Monte Carlo (MC) methods, by equilibrium molecular dynamics (EMD) simulations, and nonequilibrium molecular dynamics (NEMD) simulations. Several reviews exist for molecular simulation of diffusion in zeolites. A simulation method that allows the direct simulation of transport diffusion is the dual-control-volume grand canonical molecular dynamic (DCV-GCMD) method. Here, the chemical potential in two control volumes is kept constant by periodically performing grand canonical Monte Carlo (GCMC) insertions and deletions. BY assigning two different values to the chemical potential in the control volumes, a gradient in the chemical potential, the driving force for transport diffusion can be established. The movement of the fluid molecules is described by MD steps where Newton's equations of motion are integrated to get a physical description of the particle movement. In general, diffusion mechanisms are classified according to the interactions of the fluid molecules with the pore wall. If the pore diameter is large in comparison to the mean free path of the fluid molecules, collisions between diffusing molecules occur far more than between the molecules and the Dore walls. The diffusion mechanism is the same as in the bulk and is called molecular diffusion. DCV-GCMD simulations of counter diffusing CH_4/CF_4 mixtures in cylindrical model pores have been carried out. The diffusion mechanisms governing transport in these pores with different radii and at different temperatures. The outline of this paper is as follows. In the following section, the simulation methods as well as the fluid model and the Dore model are introduced. In the third section, the simulation results are presented and analyzed in order to determine the underlying diffusion mechanism. DCV-GCMD simulations of binary CH_4/CF_4 mixtures were carried out in the cylindrical model. Model pore used in simulations. Each pore consists of three wall layers. The wall atoms are arranged on a quadratic grid. This allowed us to continuously vary the inner pore radius (r = 13.2 A–40 A, measured from the center of the wall atoms) in a prototype system, broadly characteristic of real adsorbents, but without being constrained to, the radii of channels in known zeolite structures. The model pores consisted of three coaxial cylinders built by rings forming a quadratic arid with a distance of 1.6 A between individual wall atoms. The surface density of these model Pores was chosen to be 38 atoms/nm^2, a value close to the value of graphite (note, however, that the volume density is larger than in graphite, as the lavers were only 1.6 A apart). In DCV-GCMD simulations, the Dore is divided into two control volumes and a flow region, performing a number of GCMC insertions and deletions in the control volumes, the chemical Potential is individually controlled in each control volume and a gradient in the chemical Potential can thus be established. The standard acceptance rules for GCMC insertions and deletions are used. The movement of the molecules is described by MD moves where Newton's equation of motion re integrated in order to get the trajectory. Then the system is frozen and a new DCV-GCMD cycle starts by performing GCMC steps. It was found that in the Pores two different diffusion mechanisms occur independently surface diffusion in the layer adsorbed to the pore wall and a combination of Knudsen and molecular diffusion in the center of the Pore. Therefore, the Pores were divided in two regions: a wall region and an inner region, and the diffusivities were calculated separately in each region. In the wall region formed by the first adsorbed layer, 52 to 88% of the fluid molecules are located (the Proportion depends on the Dore radius and is larger in smaller Pores as here the influence of the Pore wall is larger) but only 5 to 19% of the flux takes Place. The diffusivities in the wall region are about two orders of magnitude smaller than in the inner region and are of the order of magnitude of liquid-phase diffusion coefficients. The simulations at different temperatures revealed that in the wall region surface diffusion–an activated Process is taking Place. The simulated activation energy of 10.7 kJ/mol for CH_4 is in the order of magnitude of experimental results. For the inner region, where only 12 to 48% of the fluid molecules are located but81 to 95% of the flux takes Place, the investigations of the radius and the temperature dependence showed that diffusion is taking place in the molecular diffusion regime or in the transition regime. For pores with radius larger than approximately 23 A, the diffusivity is no longer a function of the pore radius and diffusion is taking place in the molecular diffusion regime. The pore radius for which molecular diffusion starts to dominate is smaller for CF_4 than for CH_4 as the mean free path of CF_4 is smaller (CF_4 is the larger molecule). Our simulations demonstrate that is possible to use DCV-GCMD simulations to study in detail the complex diffusion processes that take place in different regions within the pore space of a porous solid.

1.3.3   SOME IMPORTANT SIMULATION ALGORITHMS THAT APPLIED IN PSD CALCULATION

1.3.3.1   THE MD ALGORITHM

Solving Newton's equations of motion does not immediately suggest activity at the cutting edge of research. Continuing to discuss, for simplicity, a system composed of atoms with coordinates and potential energy we introduce the atomic moment ain terms of which the kinetic energy may be written Then the energy, or Hamiltonian, may be written as a sum of kinetic and potential terms Write the classical equations of motion as. This is a system of coupled ordinary differential equations. Many methods exist to perform step-by-step numerical integration of them. Characteristics of these equations are: (a) they are 'stiff,'that is, there may be short and long timescales, and the algorithm must cope with both; (b) calculating the forces is expensive, typically involving a sum over pairs of atoms, and should be performed as infrequently as possible [157–160].

Also we must bear in mind that the advancement of the coordinates fulls two functions: (i) accurate calculation of dynamical properties, especially over times as long as typical correlation times of properties a of interest (we shall dene this later); (ii) accurately staying on the constant-energy hypersurface, for much longer times , in order to sample the correct ensemble. To ensure rapid sampling of phase space, we wish to make the time step as large as possible consistent with these requirements. For these reasons, simulation algorithms have tended to be of low order (i.e., they do not involve storing high derivatives of positions, velocities, etc.). This allows the time step to be increased as much as possible without jeopardizing energy conservation. It is unrealistic to expect the numerical method to accurately follow the true trajectory for very long time The 'ergodic' and 'mixing' properties of classical trajectories, that is, the fact that nearby trajectories diverge from each other exponentially quickly, make this impossible to achieve. All these observations tend to favor the Verlet algorithm in one form or another, and we look closely at this in the following section. For historical reasons only, we mention the more general class of predictor-corrector methods, which have been optimized for classical mechanical equations:

$$\dot{r}_i = \frac{p_i}{m_i} \, and \, \dot{p}_i = f_i \qquad (114)$$

1.3.3.2   THE VERLET ALGORITHM

There are various, essentially equivalent, versions of the Verlet algorithm, including the original method and a 'leapfrog' form. Here we concentrate on the 'velocity Verlet' algorithm, whichmay be written:

$$p_i\left(t+\frac{1}{2}\delta t\right) = p_i(t) + \frac{1}{2}\delta t f_i(t) \qquad (115)$$

$$r_i\left(t+\delta t\right)=r_i\left(t\right)+\delta t p_i\left(t+\frac{1}{2}\delta t\right)/m_i \tag{116}$$

$$p_i\left(t+\delta t\right)=p_i\left(t+\frac{1}{2}\delta t\right)+\frac{1}{2}\delta t f_i\left(t+\delta t\right) \tag{117}$$

After step (116), a force evaluation is carried out, to give for step (117). This scheme advances the coordinates and moment a over a time step. A piece of pseudocode illustrates how this works:

As we shall see shortly there is an interesting theoretical derivation of this version of the algorithm. Important features of the Verlet algorithm are: (a) it is exactly time reversible; (b) it is symplectic (to be discussed shortly); (c) it is low order in time, hencepermitting long time steps; (d) it requires just one (expensive) force evaluation per step; (e) it is easy to program.

1.3.3.3 CONSTRAINTS

It is quite common practice in classical computer simulations not to attempt to represent intra molecular bonds by terms in the potential energy function, because these bonds have very high vibration frequencies (and arguably should be treated in a quantum mechanical way rather than in the classical approximation). Instead, the bonds are treated as being constrained to have fixed length. In classical mechanics, constraints are introduced through the Lagrangian or Hamiltonian formalisms. Given an algebraic relation between two atomic coordinates, For example, a fixed bond length b between atoms 1 and 2, one may write a constraint equation, plus an equation for the time derivative of the constraint

$$\chi\left(r_1,r_2\right)=\left(r_1-r_2\right).\left(r_1-r_2\right)-b^2=0 \tag{118}$$

$$\dot{\chi}\left(r_1,r_2\right)=2\left(v_1-v_2\right).\left(r_1-r_2\right)=0 \tag{119}$$

In the Lagrangian formulation, the constraint forces acting on the atoms will enter thus:

$$m_i\ddot{r}_i=f_i+\Lambda g_i \tag{120}$$

where is the undetermined multiplier and

$$g_1=-\frac{\partial\chi}{\partial r_1}=-2\left(r_1-r_2\right)g_2=-\frac{\partial\chi}{\partial r_2}=2\left(r_1-r_2\right) \tag{121}$$

It is easy to derive an exact expression for the multiplier from the above equations; if several constraints are imposed, a system of equations (one per constraint) is obtained. However, this exact solution is not what we want: in practice, since the equations of motion are only solved approximately, in discrete time steps, the constraints will be increasingly violated as the simulation proceeds. The breakthrough in this area came with the proposal to determine the constraint forces in such a way that the constraints are satisfied exactly at the end of each time step. For the original Verlet algorithm, this scheme is called SHAKE. The appropriate version of this scheme for the velocity Verlet algorithm is called RATTLE. Formally, we wish to solve the following scheme, in which we combine $(r_1; r_2)$ into r, $(p_1; p_2)$ into etc., for simplicity:

Choosing such that: $0 = \chi\left(r(t+\delta t)\right)$ (122)

$$p(t+\delta t) = p\left(t+\frac{1}{2}\delta t\right) + \frac{1}{2}\delta t f(t+\delta t) + \mu g(t+\delta t)$$ (123)

$$p\left(t+\frac{1}{2}\delta t\right) = p(t) + \frac{1}{2}\delta t f(t+\delta t) + \lambda g(t)$$ (124)

$$r(t+\delta t) = r(t) + \frac{\delta t p\left(t+\frac{1}{2}\delta t\right)}{m}$$ (125)

Choosing such that: $0 = \chi\left(r(t+\delta t), p(t+\delta t)\right)$ (126)

Step (80) may be implemented by dening-unconstrained variables

$$\bar{p}\left(t+\frac{1}{2}\delta t\right) = p(t) + \frac{1}{2}\delta t f(t), \bar{r}(t+\delta t) = r(t) + \delta t\, \bar{p}\left(t+\frac{1}{2}\delta t\right)/m$$ (127)

then solving the non-linear equation for

$$\chi(t+\delta t) = \chi\left(\bar{r}(t+\delta t) + \lambda \delta t g(t)/m\right) = 0$$ (128)

and substituting bac

$$p\left(t+\frac{1}{2}\delta t\right) = \bar{p}\left(t+\frac{1}{2}\delta t\right) + \lambda g(t),, r(t+\delta t) = \bar{r}(t+\delta t) + \delta t \lambda g(t)/m$$ (129)

Step (81) may be handled by defining

$$\bar{p}(t+\delta t) = p\left(t+\frac{1}{2}\delta t\right) + \frac{1}{2}\delta t f(t+\delta t)$$ (130)

solving the equation for the second Lagrange multiplier

$$\dot{\chi}(t+\delta t) = \dot{\chi}\left(r(t+\delta t), \bar{p}(t+\delta t) + \mu g(t+\delta t) \right) = 0 \qquad (131)$$

(which is actually linear, since and substituting back

$$p(t+\delta t) = \bar{p}(\delta t) + \mu g(t+\delta t) \qquad (132)$$

In pseudocode this scheme may be written:

do step = 1, nstep

$$p = p + \left(\frac{dt}{2}\right) * f$$

$$r = r + dt * \frac{p}{m}$$

lambda_g = shake(r)

$$p = p + \text{lambda}_g \qquad (133)$$

$$r = r + dt * \frac{\text{lambda\_g}}{m}$$

$$f = \text{force}(r)$$

$$p = p + \left(\frac{dt}{2}\right) * f$$

mu_g = rattle(r, p)

$$p = p + \text{mu\_g}$$

enddo

The routine called shake here calculates the constraint forcesnecessary to ensure that the end-of-step positionssatisfy Eq. (114). For a system of many constraints, this calculation is usually performed in an iterative fashion, so as to satisfy each constraint in turn until convergence; the original SHAKE algorithm was framed in this way. These constraint forces are incorporated into both the end-of-step positions and the mid-step momenta. The routine called rattle calculates a new set of constraint forcesto ensure that the end-of-step momenta satisfy Eq. (115). This also may be carried out iteratively. It is important to realize that a simulation of a system with rigidly constrained bond lengths, is not equivalent to a simulation with, for example, harmonic springs representing the bonds, even in the limit of very strong springs. A subtle but crucial, difference lies in the distribution function for the other coordinates. If we obtain the conjugational distribution function by integrating over the momenta, the difference arises because in one case a set of momenta is set to zero, and not integrated, while in the other an integration is performed, whichmay lead to an extra term depending on particle coordinates. This is frequently called the metric tensor problem; it is explained in more detail in the references, and there are well-established ways of determining when the difference is likely to be significant and how to handle it, if necessary. Constraints also find an application in the study of rare events, or for convenience when it is desired to fix, for example, the director in a liquid crystal simulation. An alternative to constraints, is to retain the intramolecular bond potentials and use a multiple time step approach to handle the fast degrees of freedom [138].

1.3.3.4 ASA ALGORITHM

Althoughwhile composing each new algorithm the authors usually take into consideration the shortcomings of previous algorithms, an ultimate procedure for the solution of the linear Fredholm integral equation of the first kind has not yet been elaborated. Therefore, new efforts are needed to improve the methods for problem. First of all, the new proposed method should be the analysis of this fast, and enable one to attain approximately the same accuracy to extract f(H) from a measured global adsorption isotherm we can use genetic algorithms, evolutionary algorithms, simulated annealing, a taboo search, deterministic algorithms (our new algorithm presented in this study, the adsorption stochastic algorithm (ASA), the well-known relaxation. All of the powerful techniques mentioned above minimize all unknown variables simultaneously this algorithm can optimize the PSD calculating that obtained from mathematical modeling (Fig. 1.8).

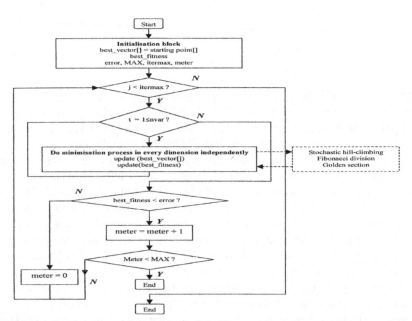

FIGURE 1.8   Flow chart of the ASA algorithm. Legend: best vector $-w$ $5(w, w,..., w)$, for the starting point a uniform distribution.

1.3.3.5 SHN ALGORITHM

Shortcomings of conventional PSD estimation techniques, as mentioned earlier, various procedures have been used conventionally to extract the pore size distributions of various adsorbents from available adsorption isotherms. In the so-called "direct methods," a prespecified isotherm was used along with some assumed distribution (for f(r)) and then the corresponding values of (Pi) were computed by algorithm. The proposed distribution was decided to be acceptable, if and only if, the computed isotherm could reconstruct the experimental data. It was clearly demonstrated in our recent article that for each local adsorption isotherm (or kernel), infinite distributions (multiple solutions) can theoretically reproduce the adsorption data, while only one of them provide proper distribution for the adsorbent. Many other suboptimal solutions (which can successfully filter-out the noise and exactly recover the true underlying isotherm hidden in a set of noisy data) can be entirely inappropriate and may lead to exceedingly misleading distributions. This is not surprising, because all of these unrealistic distributions are actually the optimal solutions of a minimization problem with no definite physical meaning. Langmuir isotherm has been employed in both synthetic isotherm data generation step and PSD recovery process. This algorithm demonstrates the optimal performances of new proposed technique (SHN2) for recovery of a pre specified true ramp function PSD from various noisy datasets using different orders of regularizations. Figure 1.9 illustrates similar performances for single, double and triple peak Gaussians as true pore size distributions. All predictions were computed using the optimum levels of regularization (*), which were found via LOOCV method and verified manually. As it can be seen in Fig.1.9he new proposed method (SHN2) provides excellent prediction for pore size distributions when appropriate order of regularization with optimum regularization level have been employed. Figure 1.9 illustrates the selected performances of our newly proposed method for various PSDs using first order regularization technique at optimum levels of regularization. In all above predictions, the Halsey correlation was used for prediction of adsorbed film thickness in both isotherm generation step and PSD recovery. In all cases, Halsey correlation was used in generation steps while Harkins and Jura (Hal-HaJu), Deboer (Hal-dB), Micromeritics (Hal-Micr), and Kruk and Jaroniec (Hal-KJ) correlations were employed,respectively, in the PSD recovery operations. Evidently, the new proposed method performs adequately for all choices of adsorbed film thickness correlations. In other words, the choice of correlation used for estimation of adsorbed film thickness is not crucial.

1.3.4 COMPARISON BETWEEN RECENT STUDY WORKS ABOUT SIMULATION AND MODELING METHODS FOR PSD CALCULATING IN CARBONS MATERIALS

Kowalczyk et al. [15] description of benzene adsorption in slit-like pores with IHK and DFT methods and by ASA algorithm and GCMC simulation. In modeling the system for PSD calculating, first BET model to determine the surface area of each sample and the quantities evaluated from adsorption. The sample cell was then placed in a liquid nitrogen bath, which created an analysis temperature of approximately 77 K with nitrogen gas used as the adsorbate. The density functional theory (DFT) was used to develop the pore size distribution data in the micropores (<2 nm) from the nitrogen isotherm with a relative pressure range from 10^{-6} to 1, while the Barrett, Joyner, Halenda (BJH) model was used for all pores greater than 2 nm. In recent works the complexity of crystallographical and geometrical structure of many solids and their complex chemical composition are the main sources of adsorbent heterogeneity. As it was pointed out by Jaroniec and Brauer the main sources of surface heterogeneity are the following: different types of crystal planes, growth steps, crystal edges and corners, various atoms and functional groups exposed at the surface and available for adsorption, irregularities in crystallographical structure of surface and impurities strongly bounded with the surface. Clearly, surface heterogeneity plays generally very important role in adsorption on solids. The process of adsorption in micropores is strongly affected by the presence of small pores of different dimensions and geometrical forms. Furthermore it is well known that uniform micropores are the source of the energetic heterogeneity in the sense that molecules adsorb in these micropores with different adsorption energies. This is caused by the change of adsorption potential field in the micropore volume. So, for microporous solids absolute heterogeneity should be identified as a superposition of non-uniformity of microporous structure and sources of, mentioned above, surface heterogeneity. At the first sight the complete description of such three-dimensional complicated structure is very difficult. Up to the present only simplified models have been developed. Clearly, different models rely on different assumptions in order to obtain relationships allowing the calculation of the main characteristics of adsorbent structural heterogeneity. Let us consider the assumptions introduced by Astinov and Do while developed the improved HK method [126, 133]:

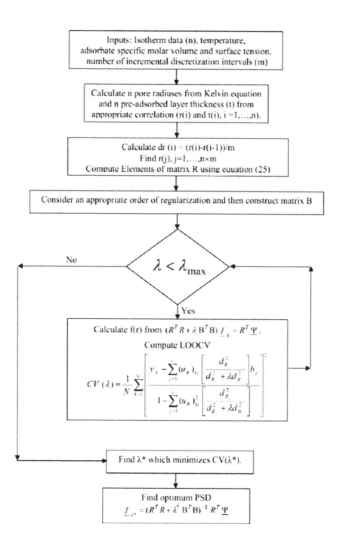

FIGURE 1.9 Flow chart for the calculation procedure of the newly proposed method (SHN2).

(1) The shape of the pores is assumed by simple slit-like. The effects of the connectivity can be neglected.

(2) The molecule–surface interaction obeys the equation of Steele.

(3) The adsorbed phase may be represented as one or two liquid-like layers between two parallel walls of a pore.

(4) The layer thickness is the distance between two parallel planes confining centers of molecules, with the density of the layer being constant over its thickness and equal to that of the corresponding liquid.

(5) The amount of the molecules of the bulk gas phase in a pore can be negligible, that is the absolute adsorption is almost equal to the surface excess adsorption.

Here, it is worth to point out the main weak points of the following assumptions. At first, the Steel's potential function seems to be very idealistic and far from real cases. It does not take under considerations all mentioned above sources of structural heterogeneity. Moreover, the integration to infinity of the potential function is also questionable. It is known that the number of graphite layers of the pore wall is limited to 3 or 4, corresponding to realistic pore wall thickness of about 1.1–1.5 nm. Such a simple model of adsorbate–adsorbent interactions, also introduced in DFT, can lead to similar drawback observed on PSD calculated from a single adsorption isotherm by the DFT method. Secondly, as pointed out by different authors, the assumption of constant density inside thin layers is the most questioned. It is known that the density varies over the layer thickness. Besides the introduced simplifications, proposed mechanism of sorption in modeled uniform silt-like pores is very similar to DFT or GCMC formalism. It is important to note here, that the final form of the equation describing the grand potential functional depends on the mechanism of adsorption. Such a mechanism is related to the size of a pore. In the very small pores having one minimum of the potential in the pore center or two symmetrical minima close to each other only one adsorbate liquid-layer mayexist. The thickness of the adsorbate layer (or adsorbed amount) depends on the relative pressure and as a consequence the local isotherm is a continuous function over a whole range of the relative pressure. In wider pores (i.e., two local minima of potential lie close the walls) the situation is more complicated. At small amount adsorbed in such type of pores two liquid layers are formed. Further increase in the relative pressure leads to the increase of the layer thickness and, consequently, to the increase of the amount adsorbed. When the distance between such two layers is small enough the mechanism of adsorption is identical to that in the small pores. So, for wider pores the first-order phase transition point should be expected.

The criterion of first-order phase transition is the equality of the grand potential in the case of double and single layer mechanism. Detailed description of mentioned above mechanism of adsorption will be elaborated later. As a result, the mechanism of pore filling is different and depends on size of pores. In narrow pores the adsorption occurs without presetting on the pore walls (i.e., single layer mechanism). In the wider pores the formation of two thin layers on the walls is followed by the first-order phase transition in the inner volume of the pore. For this reason it is necessary to consider the pore filling mechanism separately for small pores and relatively large pores. In the case of narrow pores (small micropores) the potential energy distribution with respect to distance is characterized by one minimum located at the center of the pore or two symmetrical minima near the pore walls. As a consequence, only one liquid layer mayexist in a pore. Set of equations defines the conditions of existence of a thermodynamically stable layer under the influence of an external potential field. The shape of isotherms depends on the pore width and the variation of DFs with layer thickness. This dependence should be continuous as it follows from the DFT and GCMC results. The behavior of local isotherms in small pores completely agrees with the DFT or GCMC results. The isotherms from IHK are continuous and they are shifted to the higher values of the relative pressure with increasing of effective pore widths. Effective pore width is approximated as follows, In the case of pores, which are wide enough (mesopores and some micropores), there are two local minima close to the walls. As a result, in this case, two liquid layers mayexist in the same pore. Once appeared in the pore, these two liquid layers increase with the reduced pressure and the distance between them gradually decreases. As a consequence, these two liquid like layers will coalesce into one liquid-like layer if a definite value of the reduced pressure is reached and the mechanism of pore filling changes into described above single layer. For each liquid-like layer (both layers are symmetrical) the grand potential functional is defined as follows, The IHK method seems to be a promising tool for the description of adsorption in porous solids. Since DFT and GCMC methods are insufficient for the description of benzene adsorption in micropores, IHKM seems to be the most advanced and promising procedure of porosity characterization. Presented results show that the obtained pore size distribution curves depend on the type of algorithm applied for the inverting of global adsorption isotherm equation. In spite of the fact that REG leads to the smoother PSD curves than constructed by us ASA algorithm, it can sometimes generate negative parts of distribution what is without physical meaning. On the other hand, the PSD of real heterogeneous solids is rather smooth and continuous than discreet, so regularization method used in the REG algorithm seems to be natural choice (there is necessary to take under considerations the smoothing effect on the estimated PSD). Thus, this work can state that such two algorithms in connection with intuition of researchers may be successfully applied for the proper solution of the considered ill-posed problem. The results of IHK as well as ND methods show small amount of porosity in carbon blacks applied as reference materials for the construction of as plots. The comparison of the results from IHK and ND for microporous carbons leads to the conclusion about similarity of the both methods [133].

Many researchers represented molecular models for adsorption of water on ACs and description a comparison between simulation and experimental work results. The activated carbon is modeled as being made up of non-interconnected slit pores having a distribution of pore widths given by the experimentally determined PSD. The PSD from experiment gives information on the effective, or "available" pore width w. This quantity is assumed to be related to H, the distance separating the planes through the centers of the first layer of C atoms on opposing walls, N =integral 0–infinity f(w) N(w) d(w),where f(w) is the PSD and N(w) is the amount adsorbed in the pores of width w at the given pressure. In principle, it is possible to calculate N by carrying out simulations to determine N(w) for a large number of fixed pore widths H covering the range of widths shown by the experimentally determined PSD. The total adsorption can then be determined from above integral, using the experimentally determined PSD, f(w). At low pressures a few pore widths suffice for such a scheme, since the isotherm changes slowly with pressure. For higher pressures a much larger number of pore widths is needed since pore filling occurs, and the adsorption changes rapidly. Since the simulations are lengthy, even on the fastest supercomputers, it is not feasible at present to carry out such calculations for a large number of pore widths.

In addition, in this molecular simulation method Calculations were carried out using the grand canonical Monte Carlo (GCMC) method. This method is convenient for adsorption studies, since the chemical potential í, temperature T, and volume V are specified and kept fixed in the simulation. Since í and T are the same in the bulk and adsorbed phases at equilibrium, the thermodynamic state of the bulk phase is known in such simulations. Three types of molecular moves are attempted: molecular creation, molecular destruction, and the usual Monte Carlo translation/rotation moves. Each of these three moves is attempted with equal frequency. The type of move is chosen randomly to maintain microscopic reversibility. The probability of successful creations or destructions is strongly dependent on the density of the system. The maximum allowable rotation and displacement of a molecule are adjusted so that the combined move has an acceptance probability of about 40%. This value should ensure the most efficient probing of the phase space distribution. It is noteworthy that the values of the maximum displacement and rotation are very small compared to values encountered for non-bonding systems. In these simulations the number of adsorbed molecules fluctuates during the simulation. Calculation of the average of this number for a range of chemical potential enables the adsorption isotherm to be constructed. The walls of the slit pores lie in the x–y plane. Normal periodic boundary conditions, together with the minimum image convention, are applied in these two directions. For low pressures, P/P_0 below 0.02, the length of the simulation cell in the two directions parallel to the walls was maintained at 10 nm for each of the pore widths studied, to maintain a sufficient number of adsorbed molecules. For higher pressures where more water molecules were present, the minimum cell length in the x and y directions was 4 nm. While this value is not high enough to require biased sampling methods, long runs are needed to ensure ergodicity. Associating water molecules tend to remain in energetically favorable configurations for many MC steps, and the system thus requires many steps to reach a true equilibrium state. In our runs 500 million MC steps were used for equilibration, followed by a further 500 million steps for property averaging. Shorter runs than this were not adequate to sample desorption events. The average number of water molecules in the simulation cell varied from a few molecules at the lowest pressures to a few hundred or thousands of molecules when the pores were full; filled pores contained about 320 molecules for a pore width of 0.79 nm, 460 at0.99 nm, 830 at 1.69 nm, and 2100 at 4.5 nm. Calculations were carried out on the Cornell Theory Center IBM SP2. In determining the adsorption isotherm, we commenced with the cell empty; a value of the fugacity corresponding to a low pressure was chosen and the average adsorption terminated from the simulation. The final configuration generated at each stage was used as the starting point for simulations at higher fugacities. The pressure of the bulk gas corresponding to a given chemical potential was determined from the ideal gas equation of state. Gas phase densities corresponding to the range of chemical potentials studied were determined by carrying out simulations of the bulk gas. These were found to agree with those calculated from the ideal gas equation within the estimated errors of the simulations.

Researchers reported a review about fulleren-like models for micro porous carbons that investigated structural evolution and adsorption of fulleren-like models. Modeling the structural evolution of microporous carbon including the formation mechanism of microporous carbon is not well understood at the atomic level. A number of groups have attempted to model the process, and in several cases these modeling exercises have produced structures, which contain fullerene-like elements. On the starting point for the simulation was a series of all-hexagon fragments, terminated with hydrogens, that the evolution of the structure showed that the H/Cratio is reduced (the temperature is increased). During this evolution, pentagons and heptagons form as well as hexagons, resulting in the formation of curved fragments. In each case the final carbon was made up of a hexagonal network with 10–15% non-hexagonal rings (pentagons and heptagons). The properties of the simulated carbons appeared to be generally consistent with experimental results. A different approach to modeling the evolution of microporous carbon was used. Here, the initial system consisted of carbon gas atoms at very high temperature. This choice of initial condition was intended to represent the high temperature state in a pyrolysis process after the polymer chains break down and most other elements have evaporated. The temperature was then decreased so that the atoms condensed to form a porous structure composed of curved and defected graphene sheets, in which the curvature was induced by non-hexagonal rings. In 2009, some researchers described a comprehensive molecular dynamics study of the self-assembly of carbon nanostructures. The precursor for these simulations was highly disordered amorphous carbon, which was generated by rapid quenching of an equilibrated liquid sample. It was found that, under certain conditions, annealing the amorphous carbon at high temperature could lead to the highly curved sp^2 sheet structure. In modeling adsorption using fullerene-like models for microporous carbon, there have been relatively few attempts to use fullerene-like model to predict the adsorptive and other properties of microporous carbons. By far the most ambitious program of work in this area has been carried out by Terzyk et al. [18], whose results have been published in a series of papers beginning in 2007. In the first of these 36 different carbon structures with increasing microporosity, labeled S_0–S_{35}, were generated. Fragments were then progressively added to create the 36 structures labeled S_0–S_{35}. Pore size distribution (PSD) curves for the structures were calculated using the method of Bhattacharya and Gubbins (BG). This involves determining the statistical distribution of the radii of the largest sphere that can be fitted inside a pore at a given point. It shown that the most crowded structure, S_{35}, has a much narrower range of pore sizes than the initial S_0 structure. Argon adsorption isotherms were simulated for these structures using the parallel tempering Monte Carlo simulation method. These show that the gradual crowding of the S_0 structure(leading finally to S_{35}) leads to a decrease in the maximum number of adsorbed molecules. On the other hand, the S_0 structure exhibits less adsorption at low pressures than the more crowded ones because the average micropore diameter is larger. Also notable is the increasing sharpness of the inflection point in the isotherms, a feature which is often reported for experimental systems. The simulated isotherms were then used to determine PSD curves, using a range of widely used methods, with the aim of checking the validity of these methods. Good agreement was found between the PSDs determined from the isotherms and the PSDs from the BG method. This confirms the validity of various methods for calculating PSD curves from adsorption data. It would also seem to confirm the validity of the fullerene-related model for microporous carbon. The densities of these structures were calculated, and values in the range 2.18–2.24 g cm^{-3} were found, consistent with typical densities of non-graphitizing carbons. Once again, pore size distributions for the structures were determined using the BG method. Good agreement was found between the PSDs determined from the simulated adsorption data and the original PSDs from the BG method, where the PSD curve determined from the Bhattacharya-Gubbins model. The adsorption of Ne, Ar, Kr, Xe, CCl_4 and C_6H_6 on the S_0 and S_{35} carbons was modeled. The simulated data were compared with the predictions of the Dubinin–Radushkevich and Dubinin–Astakhov adsorption isotherm equations, and a good fit was found for the S_{35} carbon. For the S_0 carbon the Dubinin–Izotova (DI) equation gave a better fit because the micropores in this model have a wide distribution of diameters. The simulated isotherms exhibited a number of features similar to those seen in experimental results. For example, the isotherms for CCl_4 and C_6H_6 were temperature invariant, as observed experimentally. It was also noted that the isotherms obeyed Gurvich's rule, which states that the larger the molecular collision diameter the smaller the access to micropores, as well as other empirical and fundamental correlations developed for adsorption on microporous carbons. The effect of oxidizing the carbon surface on porosity was analyzed in a paper published in 2009. A virtual oxidation procedure was employed, in which surface carbonyls were attached to carbon atoms located on the edges of the fragments. It was assumed that the structure of the carbon skeleton remained unchanged. Pore size distributions, determined using the BG method were found not to be greatly affected by oxidation. Simulated isotherms for Ar, N_2 and CO_2 were calculated using the GCMC method. For Ar, the effect of oxidation on the isotherm was relatively small. However, for N_2 and CO_2 there were significant changes in the isotherms, due to electrostatic interactions between N_2 and CO_2 and the surface carbonyl groups. As a consequence of this, pore size distributions calculated from the simulated isotherms for N_2 and CO_2 differed markedly from those originally determined from the BG method. An important conclusion from this is that experimental PSDs determined using CO_2 (or using N_2 if there is a large oxygen content) may be unreliable. A further study looked at the influence of carbon surface oxygen groups on Dubinin-Astakhov equation parameters calculated from CO_2 isotherms. It was concluded that porosity parameters calculated by fitting the DA model to experimental CO_2 adsorption data may be questionable. Others have published a number of other studies in which fullerene-like models have been used to predict the properties of microporous carbons, but the results summarized above are sufficient to demonstrate the utility of such models method is compared with results from the Horvath–Kawazoe method.

1.4 CONCLUSION AND OUTLOOK

This review has attempted to cover a wide range of adsorption activities of porous carbon (PC), CNTs and its composites and carbon nano structures, which have been employed so far for the removal of various pollutants from the water and wastewater. This detailed review presented more discussion about carbon nanostructures adsorbents, modeling and simulation methods in different sections. The low cost, high efficiency, simplicity and ease in the up-scaling of adsorption processes using PC makes the adsorption technique attractive for the removal and recovery of organic compounds. The activated carbon modification process has also been of interest to overcome some of the limitations of the adsorbents. Due to a large specific surface area and small, hollow, and layered structures, CNTs and carbon nano structures have been investigated as promising adsorbents for various metal ions; inorganic and organic pollutants can be easily modified by chemical treatment to increase their adsorption capacity. There is the huge hope that nanotubes applications will lead to a cleaner and healthier environment. A brief summary of these modeling methods is investigating here. In addition two important simulation method including Monte Carlo and Molecular Dynamic in this review that by design a simulation cells, dependence to materials and applied systems and other condition such parameters of models that selected for this simulation of case study, are discussed. New Monte Carlo methods will be devised and will be used with more powerful computers and applied in PSD calculating models by assumption adequate simulation cells. The presence of micro and mesopores is essential for have been many researchers aiming to control micro or mesoporosity. The present review attempted to give a general view of the recent activities on the study of pore structure control, with application novel simulation and modeling methods and necessity and important of this controlling. The aim of this review is to provide a brief overview of the methodology and modeling beside simulation methods for characterization of nanoporous carbons by using adsorption isotherm parameters. Our final goal is reported an agreed well between results as obtained by DFT and other modeling methods and our experimental works that determined from analysis of N_2 adsorption data on samples and molecular dynamic simulation or Monte Carlo simulation, that predicted optimum condition before any experimental works. The techniques the researchers employed for this purpose are unique and effective. A brief summary of these techniques is given by the carbonization of metal cation exchanged resin and the benzene CVD over ACF gave MSC with uniform microporosity or other application. For the production of mesoporous carbon, catalytic activation, polymer blend carbonization, organic gel carbonization and template carbonization methods turn out to be useful. Some of these novel methods were also applied to the formation of controlled macropores [1, 2].

All the methods introduced in this review have advantages and disadvantages. Some of them look quite unique but still far from future industrial application and some others seem to be close to practical use. However, further improvement of these methods and much effort to create a new idea for pore structure control simulation methods must be necessary to achieve the ultimate aim, that is, to prepare porous carbon with tailored pore structure. It was apparent from the literature survey that the beneficial effects of specific modification techniques on activated carbon adsorption of targeted contaminant species from aqueous solutions were profound, with some studies reported increase of contaminant uptake factors more. Concurrently, considerable decreases associated with certain contaminant uptakes can also occur depending on the technique used. The slit pore model that is currently widely used in simulating activated carbons is in part supported by physical evidence from electron microscopy, X-ray diffraction and other techniques. An alternative geometry for modeling activated carbon micropores might involve non- parallel walls. Such a wedge shaped pore model could reduce packing effects if the assumed wedge angle were large enough to blur the transitions between small integer numbers of layers of adsorbate as pore size is changed. It would be important to find independent physical evidence to support such a model [121].

A second possibility is that the explicit modeling assumption concerning the inertness of the adsorbent may not be valid. If the in vacuo pore structure of a carbon is relaxed by the relatively large quantity of adsorbate uptake at low pressure, one would expect a dilation or swelling of the structure, which in turn would reduce observable packing effects. An estimate of the driving force for pore dilation can be derived by considering a micropore of width w exposed to adsorptive at a fixed low pressure. We imagine that the pore walls resist expansion with a separation dependent potential w. We can write as a condition for dilation where n is the mols of adsorptive transferred to the pore during dilation and m is the chemical potential of the bulk adsorptive. The current widely used model for analyzing micropore distribution in activated carbon assumes an array of semiinfinite, rigid slits of distributed width whose walls are modeled as energetically uniform graphite. Adsorption isotherms can be simulated for this system using GCMC or DFT. Inversion of the integral equation of adsorption to determine micro pore size distribution from experimental isotherms using such models usually produces results showing minima near 6 and 10 A° effective pore width, regard less of the simulation method used. This is assumed to be a model induced artifact. The inclusion of surface heterogeneity in the model, while more realistic, does not change this observation significantly. The strong packing effects exhibited by a rigid parallel wall model seems likely to be the dominant feature causing the double minima in the derived pore size distributions [142].

Pore characteristics of the prepared porous samples in terms of PSD were determined using some well-known models of DS, Stoeckli, HK, IHK, DA, DFT, BET, etc. [121]. The effect of different parameters on PSD of porous carbon nano structures samples was also investigated in recent works. For example, in recent work increasing impregnation ratio increases the pore volume of activated samples, and $ZnCl_2$ resulted in ACs with more adsorption capacities than those of KOH. This increase for KOH chemical agent will create more micropores with insignificant variations of average pore size, while in the case of $ZnCl_2$ it creates wider pores, and after a specific impregnation ratio (100%) it begins to create mesopores in the carbonaceous structure. In this review three different methods: CPMD, MD and DFT is compared [129], which were adopted in nano-scale researches. CPMD, which is a typical ab initio MD, still has the difficulties in studying significantly larger systems. A novel ab initio MD method, suitable for simulating more atoms, is desirable. Classical MD allows calculations on systems containing significant numbers of atoms in a relatively long duration. However, current empirical potential functions are not accurate enough to reproduce the dynamics of molecular systems. DFT is expected to be applied in a larger system in the further. Thus, a better separation can be obtained with bigger pore sizes and relatively small distances. Therefore, an excellent separation effect can be probably obtained when varying the radius of the tube. As a result, choosing a proper temperature can greatly improve the separation. It is obvious through this review that pressures, temperatures and sorbent structures are all important factors for the separation of gas mixtures. At last, we have to stress that our simulation results depend on our choice of intermolecular potentials, but such potentials seems to be a reasonable estimate of the interactions. REBO-based MD simulations constitute a dead-end for the simulation methods. Both chief tasks of the metal catalyst can, however, be modeled successfully using DFTB/MD simulations, particularly with lower carbon feeding rates in surface diffusion simulations. CPMD simulations are too costly and cannot be employed at present for the simulation For example, for SWNT nucleation. Although we have not yet arrived at the goal of "farming" well-defined (n,m) specific SWNTs from scratch in our computers, results discussed in this review indicate that this aim is well within the reach of future DFTB/MD simulations, in particular when considering even lower carbon feedstock supply rates. Within the contents of this review we have attempted to elucidate the essential features of Monte Carlo and MD simulations and their application to problems in different modeling of PSD calculating in carbon nano adsorbents. We have attempted to give the reader practical advice as well as to present theoretically based background for the methodology of the simulations as well as the tools of analysis. In general terms we can expect that progress in Monte Carlo studies in the future will take place along two different routes. First, there will be a continued advancement towards ultra high-resolution studies of relatively simple models in which critical temperatures and exponents, adsorption condition, etc. will be examined with increasing precision and accuracy. As a consequence, high numerical resolution as well as the physical interpretation of simulation results may well provide hints to the theorist who is interested in analytic investigation. On the other hand, we expect that there will be a tendency to increase the examination of much more complicated models, which provide a better approximation to adsorption condition. As the general area of materials science blossoms, we anticipate that Monte Carlo methods will be used to probe the often-complex behavior of real materials. This is a challenge indeed, since there are usually phenomena, which are occurring at different length and time scales. As a result, it will not be surprising if multiscale methods are developed and Monte Carlo methods will be used within multiple regions of length and time scales. The general trend in Monte Carlo simulations is to ever-larger systems studied for longer and longer times. Clearly improved computer performance is moving swiftly in the direction of parallel computing. Because of the inherent complexity of message passing, it is likely that we shall see the development of hybrid computers in which large arrays of symmetric (shared memory) multiprocessors appear. (Until much higher speeds are achieved on the Internet, it is unlikely that non-local assemblies of machines will prove useful for the majority of Monte Carlo simulations.) We must continue to examine the algorithms and codes, which are used for Monte Carlo simulations to insure that they remain well suited to the available computational resources. We strongly believe that the utility of Monte Carlo simulations will continue to grow quite rapidly, but the directions may not always be predictable [1, 127–129,].

Although characterization of the energetic and geometrical heterogeneities of nanoporous carbons on the basis of gas adsorption isotherms contains a number of questions which need to be addressed in future studies, a significant progress has been done in this field in the last decades. Recent developments in sorption instrumentation allow for accurate measurements at low-pressure range for various probe molecules, which are essential for evaluation of the adsorption potential, micropore and mesopore volume distributions. The adsorption potential distribution is a model-independent function, which allows for a unique thermodynamic characterization of gas/solid adsorption systems. This distribution provides information about possible changes in the Gibbs free energy, which are caused by the energetic and geometrical heterogeneities of a nanoporous carbon as well as by the adsorbate-related entropic effects. It appears that in the case of adsorption of simple gases on nanoporous carbons their energetic heterogeneity does not change significantly the entropy effects. A general character of the adsorption potential distribution is clearly visible by its direct relation to the differential enthalpy and differential entropy. Also, the average adsorption potential is directly proportional to the heat of immersion, which through this proportionality can be estimated on the basis of vapor adsorption isotherms. Another important conclusion concerns the geometrical heterogeneity of nanoporous carbons, which is characterized by the micropore and mesopore volume distributions. The current work demonstrates that in terms of the condensation approximation both these distributions are directly related to the adsorption potential distribution. As shown the pore volume distribution can be obtained by multiplication of the adsorption potential distribution by the derivative of the adsorption potential A with respect to the pore width x. However, the pore volume distribution is a secondary characteristics of a given adsorption system because the derivative dA/dx depends on the pore geometry and adsorbate. In order to evaluate the pore volume distribution one needs to assume a model of porous structure, for example, slit-like, cylindrical or spherical pores [1, 2, 133, 145].

A brief review of methods based on the integral adsorption showed that they are attractive to evaluate the pore volume distribution. The analytical solution of this integral for subintegral functions represented by the Dubinin-Astakhov equation and gamma-type distribution is extremely simple (Jaroniec-Choma equation) and provides a good description of experimental adsorption data on nanoporous carbons. In particular, application of the gamma distribution leads to simple analytical equations for the adsorption potential distribution and other thermodynamic functions that characterize the process of the micropore filling and provide valuable information about structural and surface heterogeneities of nanoporous carbons. This description can be extended easily to adsorption of organic compounds from dilute solutions on active carbons as well as to adsorption of liquid mixtures in the whole concentration region. A significant progress has been also made in modeling adsorption in micropores. Computer simulations including Monte Carlo and MD simulation and density functional theory calculations have been recently used to evaluate the structural heterogeneity of nanoporous carbons. Methods which combine the functional density theory approach and computer simulation data with the regularization algorithm seem to be very attractive for evaluation of the pore volume distribution and energy distribution function from experimental adsorption isotherms because they do not require to assume a define shape of the distribution and are applicable to the entire range of pores. Future studies will be focused on the development and improvement of numerical methods for pore size and surface analysis. It is expected that novel carbons with uniform and ordered nanopores would play an important role in the development and examination of these methods. Another important issue in the characterization of nanoporous carbons is the elaboration of simple models based on the liquid/solid adsorption data. Although interpretation of these data is more complex, they are useful for investigation of heterogeneous nanoporous carbons.

KEYWORDS

- activated carbon fiber (ACF)
- activated carbon tailoring
- carbon nanostructures based adsorbent (CNT-CNF-ACNF, CNT-Composites)
- controlling of pore size
- modeling methods
- pore size distribution (PSD)
- simulation methods

REFERENCES

1. Bandosz, T. J., Activated carbon surfaces in environmental remediation. Vol. 7, 2006, Academic Press.
2. Marsh, H. F., Rodriguez-Reinoso, Activated carbon, 2006, Elsevier Science Limited.
3. Bach, M. T., Impact of surface chemistry on adsorption, Tailoring of activated carbon, 2007, University of Florida.
4. Machnikowski, J., et al., Tailoring Porosity Development in Monolithic Adsorbents Made of KOH-Activated Pitch Coke and Furfuryl Alcohol Binder for Methane Storage. Energy & Fuels, 2010, 24(6), 3410–3414.
5. Yin, C. Y., Aroua M. K., Daud, W. M. A. W. Review of modifications of activated carbon for enhancing contaminant uptakes from aqueous solutions: Separation and purification technology, 2007, 52(3), 403–415.
6. Muniz, J., Herrero, J., & Fuertes, A, Treatments to enhance the SO_2 capture by activated carbon fibers. Applied Catalysis B: Environmental, 1998, 18(1), 171–179.
7. Houshmand, A., W.M.A.W. Daud, & Shafeeyan, M. S., Tailoring the Surface Chemistry of Activated Carbon by Nitric Acid: Study Using Response Surface Method. Bulletin of the Chemical Society of Japan, 2011, 84(11), 1251–1260.
8. Kyotani, T., Control of pore structure in carbon: Carbon, 2000, 38(2), 269–286.
9. Kaneko, K., Determination of pore size and pore size distribution: 1. Adsorbents and catalysts. Journal of membrane science, 1994, 96(1), 59–89.

10. Olivier, J. P., Improving the models used for calculating the size distribution of micropore volume of activated carbons from adsorption data: Carbon, 1998, 36(10), 1469–1472.

11. Lastoskie, C., Gubbins, K.E., & Quirke, N., Pore size heterogeneity and the carbon slit pore: a density functional theory model. Langmuir, 1993, 9(10), 2693–2702.

12. Cao, D., & Wu, J., Modeling the selectivity of activated carbons for efficient separation of hydrogen and carbon dioxide: Carbon, 2005, 43(7), 1364–1370.

13. Feng, W., et al., Adsorption of hydrogen sulfide onto activated carbon fibers: effect of pore structure and surface chemistry. Environmental Science and Technology, 2005, 39(24), 9744–9749.

14. Banerjee, R., et al., Control of pore size and functionality in isoreticular zeolitic imidazolate frameworks and their carbon dioxide selective capture properties. Journal of the American Chemical Society, 2009,131(11), 3875–3877.

15. Kowalczyk, P., et al., Estimation of the pore size distribution function from the nitrogen adsorption isotherm. Comparison of density functional theory and the method of Do and co-workers. Carbon, 2003, 41(6), 1113–1125.

16. Tseng, R. L., & Tseng, S. K., Pore structure and adsorption performance of the KOH-activated carbons prepared from corncob. Journal of Colloid and interface Science, 2005, 287(2), 428–437.

17. Kowalczyk, P., Ciach, A., & Neimark, A. V., Adsorption-induced deformation of microporous carbons: Pore size distribution effect. Langmuir, 2008, 24(13), 6603–6608.

18. Terzyk, A. P., et al., How realistic is the pore size distribution calculated from adsorption isotherms if activated carbon is composed of fullerene-like fragments? Physical Chemistry Chemical Physics, 2007, 9(44), 5919–5927.

19. Mangun, C., et al., Effect of pore size on adsorption of hydrocarbons in phenolic-based activated carbon fibers. Carbon, 1998, 36(1), 123–129.

20. Endo, M., et al., High power electric double layer capacitor (EDLC's); from operating principle to pore size control in advanced activated carbons. Carbon science, 2001, 1 (3&4), 117–128.

21. Gadkaree, K., & Jaroniec, M., Pore structure development in activated carbon honeycombs. Carbon, 2000, 38(7), 983–993.

22. Pelekani, C., & Snoeyink, V. L. Competitive adsorption between atrazine and methylene blue on activated carbon: the importance of pore size distribution. Carbon, 2000, 38(10), 1423–1436.

23. Daud, W. M. A. W., Ali, W.S.W., & Sulaiman, M.Z. The effects of carbonization temperature on pore development in palm-shell-based activated carbon. Carbon, 2000, 38(14), 1925–1932.

24. Pelekani, C., & Snoeyink, V. L., A kinetic and equilibrium study of competitive adsorption between atrazine and Congo red dye on activated carbon: the importance of pore size distribution. Carbon, 2001, 39(1), 25–37.

25. Ebie, K., et al., Pore distribution effect of activated carbon in adsorbing organic micropollutants from natural water. Water Research, 2001, 35(1), 167–179.

26. Mangun, C. L., et al., Surface chemistry, pore sizes and adsorption properties of activated carbon fibers and precursors treated with ammonia. Carbon, 2001, 39(12), 1809–1820.

27. Moreira, R., Jose, H., & Rodrigues, A. Modification of pore size in activated carbon by polymer deposition and its effects on molecular sieve selectivity. Carbon, 2001, 39(15), 2269–2276.

28. Yang, J. B., et al., Preparation and properties of phenolic resin-based activated carbon spheres with controlled pore size distribution. Carbon, 2002, 40(6), 911–916.

29. Cao, D., et al., Determination of pore size distribution and adsorption of methane and CCl_4 on activated carbon by molecular simulation. Carbon, 2002, 40(13), 2359–2365.

30. Py, X., Guillot, A., & Cagnon, B. Activated carbon porosity tailoring by cyclic sorption/decomposition of molecular oxygen. Carbon, 2003, 41(8), 1533–1543.

31. Tanaike, O., et al., Preparation and pore control of highly mesoporous carbon from defluorinated PTFE. Carbon, 2003, 41(9), 1759–1764.

32. Zhao, J., et al., Pore structure control of mesoporous carbon as super-capacitor material. Materials Letters, 2007, 61(23), 4639–4642.

33. Chmiola, J., et al., Anomalous increase in carbon capacitance at pore sizes less than 1 nanometer. Science, 2006, 3135794, 1760–1763.

34. Lin, C., Ritter, J. A., & Popov, B. N. Correlation of Double-Layer Capacitance with the Pore Structure of Sol-Gel Derived Carbon Xerogels. Journal of the Electrochemical Society, 1999, 146(10), 3639–3643.

35. Gogotsi, Y., et al., Nanoporous carbide-derived carbon with tunable pore size. Nature materials, 2003, 2(9), 591–594.

36. Ustinov, E., Do, D., & Fenelonov, V. Pore size distribution analysis of activated carbons: Application of density functional theory using non-graphitized carbon black as a reference system. Carbon, 2006, 44(4), 653–663.

37. Han, S., et al., The effect of silica template structure on the pore structure of mesoporous carbons. Carbon, 2003, 41(5), 1049–1056.

38. Pérez-Mendoza, M., et al., Analysis of the microporous texture of a glassy carbon by adsorption measurements and Monte Carlo simulation. Evolution with chemical and physical activation. Carbon, 2006, 44(4), 638–645.

39. Dombrowski, R. J., Hyduke, D. R., & Lastoskie, C. M. Pore size analysis of activated carbons from argon and nitrogen porosimetry using density functional theory. Langmuir, 2000, 16(11), 5041–5050.

40. Khalili, N. R., et al., Production of micro and mesoporous activated carbon from paper mill sludge: I. Effect of zinc chloride activation. Carbon, 2000, 38(14), 1905–1915.

41. Dandekar, A., Baker, R., & Vannice, M., Characterization of activated carbon, graphitized carbon fibers and synthetic diamond powder using TPD and DRIFTS. Carbon, 1998, 36(12), 1821–1831.

42. Lastoskie, C., Gubbins, K. E., & Quirke, N. Pore size distribution analysis of microporous carbons: a density functional theory approach. The Journal of Physical Chemistry, 1993, 97(18), 4786–4796.

43. Kakei, K., et al., Multi-stage micropore filling mechanism of nitrogen on microporous and micrographitic carbons. J. Chem. Soc., Faraday Trans., 1990, 86(2), 371–376.

44. Sing, K. S., Adsorption methods for the characterization of porous materials. Advances in colloid and interface science, 1998, 76, 3–11.

45. Barranco, V., et al., Amorphous carbon nanofibers and their activated carbon nanofibers as super-capacitor electrodes. The Journal of Physical Chemistry C, 2010, 114(22), 10302–10307.

46. Kawabuchi, Y., et al., Chemical vapor deposition of heterocyclic compounds over active carbon fiber to control its porosity and surface function. Langmuir, 1997, 13(8), 2314–2317.

47. Miura, K., Hayashi, J. & Hashimoto, K. Production of molecular sieving carbon through carbonization of coal modified by organic additives. Carbon, 1991, 29(4), 653–660.

48. Kawabuchi, Y., et al., The modification of pore size in activated carbon fibers by chemical vapor deposition and its effects on molecular sieve selectivity. Carbon, 1998, 36(4), 377–382.

49. Verma, S. & Walker, Preparation of carbon molecular sieves by propylene pyrolysis over nickel-impregnated activated carbons. Carbon, 1993, 31(7), 1203–1207.
50. Verma, S.,. Nakayama Y., & Walker, P., Effect of temperature on oxygen-argon separation on carbon molecular sieves. Carbon, 1993, 31(3), 533–534.
51. Chen, Y., & Yang R., Preparation of carbon molecular sieve membrane and diffusion of binary mixtures in the membrane. Industrial & engineering chemistry research, 1994, 33(12), 3146–3153.
52. Rao, M. & Sircar S., Performance and pore characterization of nanoporous carbon membranes for gas separation. Journal of membrane science, 1996, 110(1), 109–118.
53. Katsaros, F., et al., High pressure gas permeability of microporous carbon membranes. Microporous materials, 1997, 8(3), 171–176.
54. Kang, I., Carbon Nanotube Smart Materials. 2005, University of Cincinnati.
55. Khan, Z. H., & Husain, M., Carbon nanotube and its possible applications. INDIAN JOURNAL OF ENGINEERING AND MATERIALS SCIENCES, 2005, 12(6), 529.
56. Khare, R., & Bose, S., Carbon nanotube based composites-a review. Journal of Minerals & Materials Characterization & Engineering, 2005, 4(1), 31–46.
57. Abuilaiwi, F. A., et al., Modification and functionalization of multi-walled carbon nanotube (MWCNT) via fischer esterification. The Arabian Journal for Science and Engineering, 2010, 35(1c), 37–48.
58. Gupta, S., & Farmer, J. Multi-walled carbon nanotubes and dispersed nanodiamond novel hybrids: Microscopic structure evolution, physical properties, and radiation resilience. Journal of Applied Physics, 2011, 109(1), 014314.
59. Upadhyayula, V. K. & Gadhamshetty, V. Appreciating the role of carbon nanotube composites in preventing biofouling and promoting biofilms on material surfaces in environmental engineering: A review. Biotechnology advances, 2010, 28(6), 802–816.
60. Saba, J., et al., Continuous electrodeposition of polypyrrole on carbon nanotube-carbon fiber hybrids as a protective treatment against nanotube dispersion, Carbon, 2012.
61. Kim, W. D., et al., Tailoring the carbon nanostructures grown on the surface of Ni–Al bimetallic nanoparticles in the gas phase. Journal of colloid and interface science, 2011, 362(2), 261–266.
62. Schwandt, C., Dimitrov, A., & Fray, D. The preparation of nano-structured carbon materials by electrolysis of molten lithium chloride at graphite electrodes. Journal of Electro analytical Chemistry, 2010,647 (2), 150–158.
63. Gao, C., et al., The new age of carbon nanotubes: An updated review of functionalized carbon nanotubes in electrochemical sensors. Nanoscale, 2012, 4(6), 1948–1963.
64. Ben-Valid, S., et al., Spectroscopic and electrochemical study of hybrids containing conductive polymers and carbon nanotubes. Carbon, 2010, 48(10), 2773–2781.
65. Vecitis, C. D., Gao, G., & Liu, H. Electrochemical carbon nanotube filter for adsorption, desorption, and oxidation of aqueous dyes and anions. The Journal of Physical Chemistry C, 2011, 115(9), 3621–3629.
66. Jagannathan, S., et al., Structure and electrochemical properties of activated polyacrylonitrile based carbon fibers containing carbon nanotubes. Journal of Power Sources, 2008,185(2), 676–684.
67. Zhu, Y., et al., Carbon-based super-capacitors produced by activation of graphene. Science, 2011, 3326037, 1537–1541.
68. Obreja, V. V., On the performance of super-capacitors with electrodes based on carbon nanotubes and carbon activated material—a review. Physica E: Low-dimensional Systems and Nanostructures, 2008, 40(7), 2596–2605.

69. Wang, L., & Yang, R. T. Hydrogen storage on carbon-based adsorbents and storage at ambient temperature by hydrogen spill over. Catalysis Reviews: Science and Engineering, 2010, 52(4), 411–461.

70. Dillon, A., et al., Carbon nanotube materials for hydrogen storage. Proceedings of the (1997) DOE/NREL Hydrogen Program Review, 1997, 237.

71. Ströbel, R., et al., Hydrogen storage by carbon materials. Journal of Power Sources, 2006, 159(2), 781–801.

72. Gupta, V. K., & Saleh, T. A., Sorption of pollutants by porous carbon, carbon nanotubes and fullerene–An overview, Environ Sci Pollut Res, 2013, 20, 2828–2843.

73. Gupta, V. K., et al., Adsorptive removal of dyes from aqueous solution onto carbon nanotubes: A review, Advances in Colloid and Interface Science, 2013, 193–194, 24–34.

74. Yang, K., et al., Competitive sorption of pyrene, phenanthrene, and naphthalene on multiwalled carbon nanotubes. Environmental science & technology, 2006, 40(18), 5804–5810.

75. Zhang, H., et al., Synthesis of a Novel Composite Imprinted Material Based on Multiwalled Carbon Nanotubes as a Selective Melamine Absorbent. Journal of agricultural and food chemistry, 2011, 59(4), 1063–1071.

76. Yan, L., et al., Characterization of magnetic guar gum-grafted carbon nanotubes and the adsorption of the dyes. Carbohydrate polymers, 2012, 87(3), 1919–1924.

77. Wu, C. H., Adsorption of reactive dye onto carbon nanotubes: equilibrium, kinetics and thermodynamics. Journal of hazardous materials, 2007, 144(1–2), 93–100.

78. Vadi, M., & Ghaseminejhad, E. Comparative Study of Isotherms Adsorption of Oleic Acid by Activated Carbon and Multi-wall Carbon Nanotube. Oriental Journal of Chemistry, 2011, 27(3), 973.

79. Vadi, M., & Moradi, N. Study of Adsorption Isotherms of Acetamide and Propionamide on Carbon Nanotube. Oriental Journal of Chemistry, 2011, 27(4), 1491.

80. Mishra, A. K., Arockiadoss, T., & Ramaprabhu, S., Study of removal of azo dye by functionalized multi walled carbon nanotubes. Chemical Engineering Journal, 2010,162(3), 1026–1034.

81. Madrakian, T., et al., Removal of some cationic dyes from aqueous solutions using magnetic-modified multi-walled carbon nanotubes. Journal of hazardous materials, 2011, 196, 109–114.

82. Kuo, C. Y., Wu, C. H., & Wu, J. Y. Adsorption of direct dyes from aqueous solutions by carbon nanotubes: Determination of equilibrium, kinetics and thermodynamics parameters. Journal of colloid and interface science, 2008, 327(2), 308–315.

83. Gong, J. L., et al., Removal of cationic dyes from aqueous solution using magnetic multiwall carbon nanotube nanocomposite as adsorbent. Journal of hazardous materials, 2009, 164(2), 1517–1522.

84. Chang, R., et al., Characterization of magnetic soluble starch-functionalized carbon nanotubes and its application for the adsorption of the dyes. Journal of hazardous materials, 2011,186 (2), 2144–2150.

85. Chatterjee, S., Lee, M. W & Woo, S.H. Adsorption of congo red by chitosan hydrogel beads impregnated with carbon nanotubes. Bioresource technology, 2010, 101(6), 1800–1806.

86. Chen, Z., et al., Adsorption behavior of epirubicin hydrochloride on carboxylated carbon nanotubes. International Journal of Pharmaceutics, 2011, 405(1), 153–161.

87. Ai, L., et al., Removal of methylene blue from aqueous solution with magnetite loaded multiwall carbon nanotube: kinetic, isotherm and mechanism analysis. Journal of hazardous materials, 2011, 198, 282–290.

88. Chronakis, I. S., Novel nanocomposites and nanoceramics based on polymer nanofibers using electro spinning process a review. Journal of Materials Processing Technology, 2005, 167(2), 283–293.
89. Inagaki, M., Yang, Y., & Kang, F., Carbon nanofibers prepared via electro spinning. Advanced Materials, 2012, 24(19), 2547–2566.
90. Im, J. S., et al., The study of controlling pore size on electrospun carbon nanofibers for hydrogen adsorption. Journal of colloid and interface science, (2008) 318(1), 42–49.
91. De Jong, K. P., & Geus, J. W. Carbon nanofibers: catalytic synthesis and applications. Catalysis Reviews, 2000, 42(4), 481–510.
92. Huang, J., Liu, Y., & You, T. Carbon nanofiber based electrochemical biosensors: A review. Analytical Methods, 2010, 2(3), 202–211.
93. Yusof, N., & Ismail, A. Post spinning and pyrolysis processes of polyacrylonitrile (PAN)-based carbon fiber and activated carbon fiber: A review. Journal of Analytical and Applied Pyrolysis, 2012, 93, 1–13.
94. Sullivan, P., et al., Physical and chemical properties of PAN-derived electrospun activated carbon nanofibers and their potential for use as an adsorbent for toxic industrial chemicals. Adsorption, 2012, 18(3–4), 265–274.
95. Tavanai, H., Jalili, R., & Morshed, M. Effects of fiber diameter and CO_2 activation temperature on the pore characteristics of polyacrylonitrile based activated carbon nanofibers. Surface and Interface Analysis, 2009, 41(10), 814–819.
96. Wang, G., et al., Activated carbon nanofiber webs made by electro spinning for capacitive deionization. Electrochimica Acta, 2012, 69, 65–70.
97. Ra, E. J., et al., Ultramicropore formation in PAN/camphor-based carbon nanofiber paper. Chemical Communications, 2010, 46(8), 1320–1322.
98. Liu, W., & Adanur, S., Properties of electrospun polyacrylonitrile membranes and chemically activated carbon nanofibers. Textile Research Journal, 2010, 80(2), 124–134.
99. Korovchenko, P., Renken, A., & Kiwi-Minsker, L. Microwave plasma assisted preparation of Pd-nanoparticles with controlled dispersion on woven activated carbon fibers. Catalysis today, 2005, 102, 133–141.
100. Jung, K. H., & Ferraris, J. P. Preparation and electrochemical properties of carbon nanofibers derived from polybenzimidazole/polyimide precursor blends. Carbon, 2012.
101. Im, J. S., Park, S. J., & Lee, Y. S. Preparation and characteristics of electrospun activated carbon materials having meso-and macropores. Journal of colloid and interface science, 2007, 314(1), 32–37.
102. Esrafilzadeh, D., Morshed, M., & Tavanai, H., An investigation on the stabilization of special polyacrylonitrile nanofibers as carbon or activated carbon nanofiber precursor. Synthetic Metals, 2009, 159(3), 267–272.
103. Hung, C. M., Activity of Cu-activated carbon fiber catalyst in wet oxidation of ammonia solu tion. Journal of hazardous materials, 2009, 166(2), 1314–1320.
104. Koslow, E. E., Carbon or activated carbon nanofibers, 2007, Google Patents.
105. Lee, J. W., et al., Heterogeneous adsorption of activated carbon nanofibers synthesized by electro spinning polyacrylonitrile solution. Journal of nanoscience and nanotechnology, 2006, 6(11), 3577–3582.
106. Oh, G. Y., et al., Preparation of the novel manganese-embedded PAN-based activated carbon nanofibers by electro spinning and their toluene adsorption. Journal of Analytical and Applied Pyrolysis, 2008, 81(2), 211–217.
107. Zussman, E., et al., Mechanical and structural characterization of electrospun PAN-derived carbon nanofibers. Carbon, 2005, 43(10), 2175–2185.

108. Kim, C., Electrochemical characterization of electrospun activated carbon nanofibers as an electrode in super-capacitors. Journal of Power Sources, 2005, 142(1), 382–388.

109. Jung, M. J., et al., Influence of the textual properties of activated carbon nanofibers on the performance of electric double-layer capacitors. Journal of Industrial and Engineering Chemistry, 2013.

110. Fan, Z., et al., Asymmetric super-capacitors based on graphene/MnO$_2$ and activated carbon nanofiber electrodes with high power and energy density. Advanced Functional Materials, 2011, 21(12), 2366–2375.

111. Jeong, E., Jung, M. J., & Lee, Y. S Role of fluorination in improvement of the electrochemical properties of activated carbon nanofiber electrodes. Journal of Fluorine Chemistry, 2013.

112. Seo, M. K., & Park, S. J., Electrochemical characteristics of activated carbon nanofiber electrodes for super-capacitors. Materials Science and Engineering: B, 2009, 164(2), 106–111.

113. Endo, M., et al., High power electric double layer capacitor (EDLC's); from operating principle to pore size control in advanced activated carbons. Carbon science, 2001, 1(3&4), 117–128.

114. Ji, L., & Zhang, X. Generation of activated carbon nanofibers from electrospun polyacrylonitrile-zinc chloride composites for use as anodes in lithium-ion batteries. Electrochemistry Communications, 2009, 11(3), 684–687.

115. Karra, U., et al., Power generation and organics removal from wastewater using activated carbon nanofiber (ACNF) microbial fuel cells (MFCs). International Journal of Hydrogen Energy, 2012.

116. Oh, G. Y., et al., Adsorption of toluene on carbon nanofibers prepared by electro spinning. Science of the Total Environment, 2008, 393(2), 341–347.

117. Lee, K. J., et al., Activated carbon nanofiber produced from electrospun polyacrylonitrile nanofiber as a highly efficient formaldehyde adsorbent. Carbon, 2010, 48(15), 4248–4255.

118. Katepalli, H., et al., Synthesis of hierarchical fabrics by electro spinning of PAN nanofibers on activated carbon microfibers for environmental remediation applications. Chemical Engineering Journal, 2011, 171(3), 1194–1200.

119. Gaur, V., Sharma, A. & Verma, N. Preparation and characterization of ACF for the adsorption of BTX and SO$_2$, Chemical Engineering and Processing: Process Intensification, 2006, 45(1), 1–13.

120. Cheng, K. K., Hsu, T. C., & Kao, L. H. A microscopic view of chemically activated amorphous carbon nanofibers prepared from core/sheath melt-spinning of phenol formaldehyde-based polymer blends. Journal of materials science, 2011, 46(11), 3914–3922.

121. Zhijun, et al., A Molecular Simulation Probing of Structure and Interaction for Supramolecular Sodium Dodecyl Sulfate/S-W Carbon Nanotube Assemblies. Nano Lett, 2009.

122. Ong, et al., Molecular Dynamics Simulation of Thermal Boundary Conductance Between Carbon Nanotubes and SiO$_2$, Phys. Rev. B, 81, 2010.

123. Yang, et al., Preparation and properties of phenolic resin-based activated carbon spheres with controlled pore size distribution, Carbon, 2002, 40(5), 911–916.

124. Liu, et al., Carbon Nanotube Based Artificial Water Channel Protein: Membrane Perturbation and Water Transportation, Nano Lett, 2009, 9 (4), 1386–1394.

125. Zhang, et al., Interfacial Characteristics of Carbon Nanotube-Polyethylene Composites Using Molecular Dynamics Simulations, ISRN Materials Science, Article ID 145042, 2011.

126. Mangun, L., et al., Surface chemistry, pore sizes and adsorption properties of activated carbon fibers and precursors treated with ammonia, Carbon, 2001, 139(11), 1809–1820.
127. Irle, et al., Milestones in Molecular Dynamics Simulations of Single-Walled Carbon Nanotube Formation: A Brief Critical Review, Nano Res, 2009, 2(8), 755–767.
128. Guo-zhuo, et al., Regulation of pore size distribution in coal-based activated carbon, New Carbon Materials, 2009, 24(2), article ID: 1007–8827.
129. Zang, et al., A comparative study of Young's modulus of single-walled carbon nanotube by CPMD, MD and first principle simulations, Computational Materials Science, 2009, 46(4), 621–625.
130. Frankland, et al., Molecular Simulation of the Influence of Chemical Cross-Links on the Shear Strength of Carbon Nanotube-Polymer Interfaces, J. Phys. Chem. B., 2002, 106(2), 3046–3048.
131. Frankland, et al., Simulation for separation of hydrogen and carbon monoxide by adsorption on single-walled carbon nanotubes, Fluid Phase Equilibria, 2002, 194–197(10), 297–307.
132. Han, et al., Molecular dynamics simulations of the elastic properties of polymer/carbon nanotube composites, Computational Materials Science, 2007, 39(8), 315–323.
133. Okhovat, et al., Pore Size Distribution Analysis of Coal-Based Activated Carbons: Investigating the Effects of Activating Agent and Chemical Ratio, ISRN Chemical Engineering, 2012.
134. Lehtinen, et al., Effects of ion bombardment on a two-dimensional target: Atomistic simulations of graphene irradiation, Physical Review B, 2010, 81.
135. Shigeo Maruyama, A Molecular Dynamics Simulation of Heat Conduction of a Finite Length Single-Walled Carbon Nanotube, Microscale Thermophysical Engineering, 2003, 7, 41–50.
136. Williams, et al., Monte Carlo simulations of H_2 physisorption in finite-diameter carbon nanotube ropes, Chemical Physics Letters, 2000, 320(6), 352–358.
137. Diao, et al., Molecular dynamics simulations of carbon nanotube/silicon interfacial thermal conductance, The Journal Of Chemical Physics, 2008, 128.
138. Noel, et al., On the Use of Symmetry in the Ab Initio Quantum Mechanical Simulation of Nanotubes and Related Materials, J Comput Chem, 2010, 31, 855–862.
139. Nicholls, D. et al., Water Transport Through (7,7) Carbon Nanotubes of Different Lengths using Molecular Dynamics, Microfluidics and Nanofluidics, 2012, 1–4, 257–264.
140. Ribas, A. et al., Nanotube nucleation versus carbon-catalyst adhesion–Probed by molecular dynamics simulations, J. Chem. Phys, 2009, 131.
141. Sanz-Navarro, F., et al., Molecular Dynamics Simulations of Metal Clusters Supported on Fishbone Carbon Nanofibers, J. Phys. Chem. C, 2010, 114, 3522–3530.
142. Shibuta, et al., Bond-order potential for transition metal carbide cluster for the growth simulation of a single-walled carbon nanotube Department of Materials Engineering, The University of Tokyo, 2003.
143. Mao, et al., Molecular dynamics simulations of the filling and decorating of carbon nanotubules, Nanotechnology, 1999, 10, 273–277.
144. Mao, et al., Molecular Simulation Study of CH_4/H_2 Mixture Separations Using Metal Organic Framework Membranes and Composites, J. Phys. Chem. C, 2010, 114, 13047–13054.
145. Thomas, A., et al., Pressure-driven Water Flow through Carbon Nanotubes: Insights from Molecular Dynamics Simulation, Carnegie Mellon University, USA, Department of Mechanical Engineering, 2009.

APPENDIX A

PROGRAM 1: TEST A RANDOM NUMBER GENERATOR IN MONTE CARLO SIMULATION

Note, as an exercise the student may wish to insert other random number generators or add tests to this simple program.

****

This program is used to perform a few very simple tests of a random number generator. A congruential generator is being tested

****

```
Real*8 Rnum(100,000),Rave,R2Ave,Correl,SDev
Integer Iseed,num
open(Unit=1,®le='result_testrng_02')
PMod = 2147483647.0D0
DMax = 1.0D0/PMod
*******
Input
*******
write(*,800)
800     format('enter the random number generator seed ')
read(*,921) Iseed
921     format(i5)
write(*,801) Iseed
write(1,801) Iseed
801     format(' The random number seed is ', I8)
write(*,802)
802     format('enter the number of random numbers to be generated')
read(*,921) num
write(*,803) num
write(1,803)num
803     format ('number of random numbers to be generated = ',i8)
****************************
Initialize variables, vectors
****************************
do 1 i=2,10000
    Rnum(i)=0.0D0
Rave=0.D0
Correl=0.0D0
R2Ave=0.0D0
SDev=0.0D0
************************
```

Calculate random numbers
****
```
Rnum(1)=Iseed*DMax
Write(*,931) Rnum(1),Iseed
Do 10 i=2,num
Rnum(i)=cong16807(Iseed)
if (num.le.100) write(*,931) Rnum(i),Iseed
931     format(f10.5,i15)
10 continue
Rave=Rnum(1)
R2Ave=Rnum(1)**2
Do 20 i=2,num
Correl=Correl+Rnum(i)*Rnum(i–1)
Rave=Rave+Rnum(i)
20 R2Ave=R2Ave+Rnum(i)**2
Rave=Rave/num
SDev=Sqrt((R2Ave/num-Rave**2)/(num–1))
Correl=Correl/(num–1)-Rave*RAve
```
****
Output
****
```
write(*,932) Rave,SDev,Correl
932     format('Ave. random number =',F10.6, ' +/–,' F10.6,
1/'nn-correlation =' F10.6)
write(1,932) Rave,SDev,Correl
999     format(f12.8)
close (1)
stop
end
```

```
FUNCTION Cong16807(ISeed)
```
****
This is a simple congruential random number generator
****
```
INTEGER ISeed,IMod
REAL*8 RMod,PMod,DMax
RMod = DBLE(ISeed)
PMod = 2147483647.0D0
DMax = 1.0D0/PMod
RMod = RMod*16807.0D0
IMod = RMod*DMax
RMod = RMod–PMod*IMod
```

```
cong16807=rmod*DMax
Iseed=Rmod
RETURN
END
```

PROGRAM 2: A GOOD ROUTINE FOR GENERATING A TABLEOF RANDOM NUMBERS IN MONTE CARLO SIMULATION.

```
**************************************************************
```
This program uses the R250/R521 combined generator described in:
A. Heuer, B. Duenweg and A.M. Ferrenberg, Comp. Phys. Comm. 103, 1
1997). It generates a vector, RanVec, of length RanSize 31-bit random
integers. Multiply by RMaxI to get normalized random numbers. You
will need to test whether RanCnt will exceed RanSize. If so, call
GenRan again to generate a new block of RanSize numbers. Always
remember to increment RanCnt when you use a number from the table.
```
**************************************************************
IMPLICIT non-E
INTEGER RanSize,Seed,I,RanCnt,RanMax
PARAMETER(RanSize = 10000)
PARAMETER(RanMax = 2147483647)
INTEGER RanVec(RanSize),Z1(250+RanSize),Z2(521+RanSize)
REAL*8 RMaxI
PARAMETER (RMaxI = 1.0D0/(1.0D0*RanMax))
COMMON/MyRan/RanVec,Z1,Z2,RanCnt
SAVE
Seed = 432987111
*****************************************
```
Initialize the random number generator.
```
*****************************************
CALL InitRan(Seed)*
**************************************************************
```
If the 10 numbers we need pushes us past the end of the RanVec vector,
call GenRan. Since we just called InitRan, RanCnt = RanSize we must
call it here.
```
**************************************************************
IF ((RanCnt + 10).GT. RanSize) THEN
    Generate RanSize numbers and reset the RanCnt counter to 1
Call GenRan
END IF
Do I = 1,10
WRITE(*,*) RanVec(RanCnt + I–1),RMaxI*RanVec(RanCnt + I–1)
End Do
```

```
Rancnt = Rancnt + 10
*****************************************************************
Check to see if the 10 numbers we need will push us past the end
of the RanVec vector. If so, call GenRan.
*****************************************************************
IF ((RanCnt + 10).GT. RanSize) THEN
    Generate RanSize numbers and reset the RanCnt counter to 1
Call GenRan
END IF
Do I = 1,10
WRITE(*,*) RanVec(RanCnt + I–1),RMaxI*RanVec(RanCnt + I–1)
End Do
RanCnt = RanCnt + 10
END
SUBROUTINE InitRan(Seed)
*****************************************************************
Initialize the R250 and R521 generators using a congruential generator
to set the individual bits in the 250/521 numbers in the table. The
R250 and R521 are then warmed-up by generating 1000 numbers.
*****************************************************************
IMPLICIT non-E
REAL*8 RMaxI,RMod,PMod
INTEGER RanMax,RanSize
PARAMETER(RanMax = 2147483647)
PARAMETER(RanSize = 100,000)
PARAMETER(RMaxI = 1.0D0/(1.0D0*RanMax))
INTEGER Seed,I,J,K,IMod,IBit
INTEGER RanVec(RanSize),Z1(250+RanSize),Z2(521+RanSize)
INTEGER RanCnt
COMMON/MyRan/RanVec,Z1,Z2,RanCnt
SAVE
RMod = DBLE(Seed)
PMod = DBLE(RanMax)
***********************************
Warm up a congruential generator
***********************************
Do I = 1,1000
RMod = RMod*16807.0D0
IMod = RMod/PMod
RMod = RMod–PMod*IMod
End Do
*****************************************************************
```

Now fill up the tables for the R250 & R521 generators: This requires random integers in the range 0±> 2*31 1. Iterate a strange number of times to improve randomness.

```
*************************************************************
Do I = 1,250
Z1(I) = 0
IBit = 1
Do J = 0,30
Do K = 1,37
RMod = RMod*16807.0D0
IMod = RMod/PMod
RMod = RMod – PMod*IMod
End Do
    Now use this random number to set bit J of X(I).
IF (RMod. GT. 0.5D0*PMod) Z1(I) = IEOR(Z1(I),IBit)
IBit = IBit*2
End Do
End Do
Do I = 1,521
Z2(I) = 0
IBit = 1
Do J = 0,30
Do K = 1,37
RMod = RMod*16807.0D0
IMod = RMod/PMod
RMod = RMod–PMod*IMod
End Do
    Now use this random number to set bit J of X(I).
IF (RMod. GT. 0.5D0*PMod) Z2(I) = IEOR(Z2(I),IBit)
IBit = IBit*2
End Do
End Do
*************************************************************
```

Perform a few iterations of the R250 and R521 random number generators to eliminate any effects due to'poor' initialization.

```
*************************************************************
Do I = 1,1000
Z1(I+250) = IEOR(Z1(I),Z1(I+147))
Z2(I+521) = IEOR(Z2(I),Z2(I+353))
End Do
Do I = 1,250
Z1(I) = Z1(I + 1000)
```

```
End Do
Do I = 1,521
Z2(I) = Z2(I + 1000)
End Do
*************************************************************
Set the random number counter to RanSize so that a proper checking
code will force a call to GenRan in the main program.
*************************************************************
RanCnt = RanSize
RETURN
END


SUBROUTINE GenRan
*************************************************************
Generate vector RanVec (length RanSize) of pseudorandom 31-bit
integers.
*************************************************************
IMPLICIT non-E
INTEGER RanSize,RanCnt,I
PARAMETER(RanSize = 100000)
INTEGER RanVec(RanSize),Z1(250+RanSize),Z2(521+RanSize)
COMMON/MyRan/RanVec,Z1,Z2,RanCnt
SAVE
*************************************************************
Generate RanSize pseudorandom numbers using the individual generators
*************************************************************
Do I = 1,RanSize
Z1(I+250) = IEOR(Z1(I),Z1(I+147))
Z2(I+521) = IEOR(Z2(I),Z2(I+353))
End Do
*************************************************************
Combine the R250 and R521 numbers and put the result into RanVec
*************************************************************
Do I = 1,RanSize
RanVec(I) = IEOR(Z1(I+250),Z2(I+521))
End Do
*************************************************************
Copy the last 250 numbers generated by R250 and the last 521 numbers
from R521 into the working vectors (Z1), (Z2) for the next pass.
*************************************************************
Do I = 1,250
Z1(I) = Z1(I + RanSize)
```

```
End Do
Do I = 1,521
Z2(I) = Z2(I + RanSize)
End Do
****************************************
Reset the random number counter to 1.
****************************************
RanCnt = 1
RETURN
END
```

CHAPTER 2

MICRO AND NANOSCALE SYSTEMS

CONTENTS

2.1 INTRODUCTION

Viscous forces in the fluid can lead to large dispersion flow along the axis of motion. They have a significant impact, both on the scale of individual molecules, and the scale of microflows; near the borders of the liquid-solid (beyond a few molecular layers), during the motion on a complex and heterogeneous borders.

Influence of the effect of boundary regions on the particles and fluxes have been observed experimentally in the range of molecular thicknesses up to hundreds of nanometers. If the surface has a super hydrophobic property, this range can extend to the micron thickness. *Molecular theory can predict the effect of hydrophobic surfaces in the system only up to tens of nanometers.*

Fluids, the flow of liquid or gas, have properties that vary continuously under the action of external forces. In the presence of fluid shear forces are small in magnitude, leads large changes in the relative position of the element of fluid. In contrast, changes in the relative positions of atoms in solids remain small under the action of any small external force. Termination of action of the external forces on the fluid does not necessarily lead to the restoration of its initial form.

2.1.1 CAPILLARY EFFECTS

To observe the capillary effects, one must open the nanotube, that is, to remove the upper part lids. Fortunately, this operation is quite simple.

The first study of capillary phenomena has shown that there is a relationship between the magnitude of surface tension and the possibility of its being drawn into the channel of the nanotube. It was found that the liquid penetrates into the channel of the nanotube, if its surface tension is not higher than 200 mN/m. For example, concentrated nitric acid with surface tension of 43 mN/m is used to inject certain metals into the channel of a nanotube. Then annealing is conducted at 4000°C for 4 h in an atmosphere of hydrogen, which leads to the recovery of the metal.

Along with the metals, carbon nanotubes can be filled with gaseous substances, such as hydrogen in molecular form. This ability is of great practical importance and can be used as a clean fuel in internal combustion engines.

2.1.2 SPECIFIC ELECTRICAL RESISTANCE OF CARBON NANOTUBES (R)

The resistivity of the nanotubes can be varied within wide limits to 0.8 ohm/cm. The minimum value is lower than that of graphite. Most of the nanotubes have metallic conductivity, and the smaller shows properties of a semiconductor with a band gap of 0.1 to 0.3 eV.

The resistance of single-walled nanotube is independent of its length, because of this it is convenient to use for the connection of logic elements in microelectronic devices. The permissible current density in carbon nanotubes is much greater than in metallic wires of the same cross section and one hundred times better achievement for superconductors.

2.1.3   EMISSION PROPERTIES OF CARBON NANOTUBES

The results of the study of emission properties of the material (where the nano-tubes were oriented perpendicular to the substrate) have been very interesting for practical use. An attained value of the emission current density is of the order of 0.5 mA/mm^2. The value obtained is in good agreement with the Fowler-Nordheim expression.

2.2   ELECTRO-KINETIC PROCESSES IN MICRO AND NANOSCALE SYSTEMS

The most effective and common ways to control microflow substances are *electrokinetic* and *hydraulic*. At the same time the most technologically advanced and automated considered electrokinetic.

Charges transfer in mixtures occurs as a result of the directed motion of charge carriers' ions. There are different mechanisms of such transfer, but usually are *convection, migration and diffusion*.

Convection is called mass transfer the macroscopic flow. *Migration* is the movement of charged particles by electrostatic fields. The velocity of the ions depends on field strength. In microfluidics the *electrokinetic process* can be divided into: *electro-osmosis, electrophoresis, streaming potential* and *sedimentation potential*. These processes can be qualitatively described as follows:

a) *electro-osmosis*; the movement of the fluid volume in response to the applied electric field in the channel of the electrical double layers on its wetted surfaces.

b) *Electrophoresis*; the forced motion of charged particles or molecules, in mixture with the acting electric field.

c) *Streamy potential*; the electric potential, which is distributed through a channel with charged walls, in the case when the fluid moves under the action of pressure forces. Joule electric current associated with the effect of charge transfer is flowing stream.

d) *The potential of sedimentation*; an electric potential is created when charged particles are in motion relative to a constant fluid.

In general, for the microchannel cross-section S amount of introduced probe (when entering electrokinetic method) depends on the applied voltage U, time t during which the received power, and mobility of the sample componentsm:

$$Q = \frac{\mu S U t}{L} \cdot c$$

where c – probeconcentrationin the mixture; L–the channel length.

Amount of injected substance is determined by the electrophoretic and total electro-osmotic mobilitiesm.

In the hydrodynamic mode of entry by the pressure difference in the channel or capillary of circular cross section, the volume of injected probe V_c:

$$V_c = \frac{4}{128} \cdot \frac{\Delta p \pi dt}{\eta L}$$

where Δp–pressure differential; d–diameter of the channel; h– viscosity.

2.3 CONTINUUM HYPOTHESIS

In the simulation of processes in micron-sized systems the following basic principles are fundamentals:

1) Hypothesis of *laminar* flow (sometimes is taken for granted when it comes to microfluidics);
2) Continuum hypothesis (detection limits of applicability);
3) Laws of formation of the velocity profile, mass transfer, the distribution of electric and thermal fields;
4) Boundary conditions associated with the geometry of structural elements (walls of channels, mixers zone flows, etc.).

Since we consider the physical and chemical transport processes of matter and energy, mathematical models, have the form of systems of differential equations of second order partial derivatives. Methods for solving such equations are analytical (Fourier and its modifications, such as the method of Greenberg, Galerkin, the method of d'Alembert and the Green's functions, the Laplace operator method, etc.) or numerical (explicit or more effectively, implicit finite difference schemes).

Laminar flow; a condition in which the particle velocity in the liquid flow is not a random function of time. The small size of the microchannels (typical dimensions of 5 to 300 microns) and low surface roughness create good conditions for the establishment of laminar flow. Traditionally, the image of the nature of the flow gives the dimensionless characteristic numbers: the Reynolds number and Darcy's friction factor.

In the motion of fluids in channels the turbulent regime is rarely achieved. At the same time, the movement of gases is usually turbulent.

Although the liquids are quantized in the length scale of intermolecular distances (about 0.3 nm to 3 nm in liquids and for gases), they are assumed to be continuous in most cases. Continuum hypothesis (continuity, continuum) suggests that the macroscopic properties of fluids consisting of molecules, the same as if the fluid were completely continuous (structurally homogeneous). Physical characteristics: mass, momentum and energy associated with the volume of fluid containing a sufficiently large number of molecules must be taken as the sum of all the relevant characteristics of the molecules.

Continuum hypothesis leads to the concept of fluid particles. In contrast to the ideal of a point particle in ordinary mechanics, in fluid mechanics, particle in the fluid has a finite size.

At the atomic scale there are large fluctuations due to the molecular structure of fluids, but if we the increase the sample size, we reach a level where it is possible to obtain stable measurements. This volume of probe must contain a sufficiently large number of molecules to obtain reliable reproducible signal with small statistical fluctuations. For example, if we determine the required volume as a cube with sides of 10 nm, this volume contains some of the molecules and determines the level of fluctuations of the order of 0.5%.

The applicability of the hypothesis is based on comparison of free path length of a particle λ in a liquid with a characteristic geometric size d. The ratio of these lengths called the Knudsen number: $Kn = \lambda/d$.

(a) When $Kn < 10^{-3}$ justifies hypothesis of a continuous medium, and (b) when $Kn < 10^{-1}$ allows the use of adhesion of particles to the solid walls of the channel. Theoretical condition can also be varied: both in form $U = 0$ and in a more complex form, associated with shear stresses. The calculation of λ can be carried out as $\lambda \approx \sqrt[3]{\overline{V}/Na}$,

where, $\overline{V}$ – molar volume; Na – Avogadro's number.

Under certain geometrical approximations of the particles of substance, free path length can be calculated as $\lambda = 1/(\sqrt{2}\pi r_s^2 Na)$, (if used instead r_s Stokes radius, as a consequence of the spherical approximation of the particle). On the other hand, for a rigid model of the molecule r_s should be replaced by the characteristic size of the particles R_g (the radius of inertia), calculated as $R_g = n_i.\delta_l / \sqrt{6}$, where, δ_l – the length of a fragment of the chain (link); n_l – the number of links.

Of course, the continuum hypothesis is not acceptable when the system under consideration is close to the molecular scale. This happens in nanoliquid, such as liquid transport through *nano*-pores in cell membranes or artificially made nano-channels.

2.4 THE MOLECULAR DYNAMICS METHOD

In contrast to the continuum hypothesis, the essence of modeling the molecular dynamics method is as follows. We consider a large ensemble of particles which simulate atoms or molecules, that is, all atoms are material points. It is believed that the particles interact with each other and, moreover, may be subject to external influence. Inter-atomic forces are represented in the form of the classical potential force (the gradient of the potential energy of the system).

The interaction between atoms is described by means of van der Waals forces (intermolecular forces), mathematically expressed by the Lennard-Jones potential:

$$V(r) = \frac{Ae^{-\sigma r}}{r} - \frac{C_6}{r^6}$$

where, A and C_6– some coefficients depending on the structure of the atom or molecule;s – the smallest possible distance between the molecules.

In the case of two isolated molecules at a distance of r_0 the interaction force is zero (that is, the repulsive forces balance attractive forces). When $r > r_0$ the resultant force is the force of gravity, which increases in magnitude, reaching a maximum at $r = r_m$ and then decreases. When $r > r_0$ there is a repulsive force. Molecule in the field of these forces has potential energy $V(r)$, which is connected with the force of $f(r)$ by the differential equation

$$dv = -f(r)dr .$$

At the point where $r = r_0$, $f(r) = 0$, $V(r)$reachesan extremum (minimum).

The chart of such a potential is shown below in Fig. 2.1. The upper (positive) half-axis r corresponds to the repulsion of the molecules, the lower (negative) half-plane shows their attraction. We can now observe that at short distances the molecules mainly repel each other.

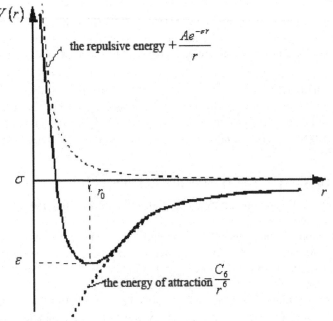

FIGURE 2.1   Potential energy of intermolecular interaction.

The exponential summand in the expression for the potential describing the repulsion of the molecules at small distances, often approximated as

$$\frac{Ae^{-\sigma r}}{r} \approx \frac{C_{12}}{r^{12}}$$

In this case we obtain the Lennard-Jones potential:

$$V(r) = \frac{C_{12}}{r^{12}} - \frac{C_6}{r^6} \tag{2.1}$$

The interaction between carbon atoms is described by the potential

$$V_{CC}(r) = K(r-b)^2,$$

where, $K = 326\text{Дж}/\text{м}^2$ – constant tension (compression) connection; $b = 1,4A$ – the equilibrium length of connection; r – current length of the connection.

The interaction between the carbon atom and hydrogen molecule is described by the Lennard-Jones

$$V(r) = 4\varepsilon\left[\left(\frac{\sigma}{r}\right)^{12} - \left(\frac{\sigma}{r}\right)^6\right]$$

For all particles (Fig. 2.2), the equations of motion are written:

$$m\frac{d^2\overline{r_i}}{dt^2} = \overline{F}_{T-H_2}\left(\overline{r_i}\right) + \sum_{j\neq i}\overline{F}_{H_2-H_2}\left(\overline{r_i} - \overline{r_j}\right),$$

where, $\overline{F}_{T-H_2}(\overline{r})$ – Force, acting by the CNT.

The resulting system of equations is solved numerically. However, the molecular dynamics method has limitations of applicability:
1) the de Broglie wavelength h/mv(where, h – Planck's constant;m– the mass of the particle; v–velocity);
2) Classical molecular dynamics cannot be applied for modeling systems consisting of light atoms such as helium or hydrogen;
3) At low temperatures, quantum effects become decisive for the consideration of such systems must use quantum chemical methods;
4) The time at which we consider the behavior of the system were more than the relaxation time of the physical quantities.

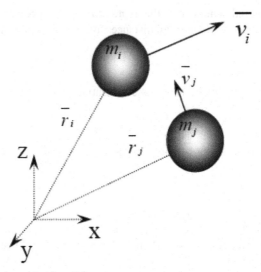

FIGURE 2.2   Schematic of particles.

$\overline{\Gamma}_{H_2-H_2}(\overline{r}_i - \overline{r}_j)$ – Force acting on the i-th molecule from the j-th molecule.

The coordinates of the molecules are distributed regularly in the space. The velocities of the molecules are distributed according to the Maxwell equilibrium distribution function according to the temperature of the system:

$$f(u,v,w) = \frac{\beta^3}{\pi^{3/2}} \exp\left(-\beta^2\left(u^2 + v^2 + w^2\right)\right), \beta = \frac{1}{\sqrt{2RT}}$$

The macroscopic flow parameters are calculated from the distribution of positions and velocities of the molecules:

$$\overline{V} = \overline{v}_i = \frac{1}{n}\sum_i \overline{v}_i$$

$$\rho = \frac{nm}{V_0}$$

$$\frac{3}{2}RT = \frac{1}{2}\left|\overline{v}_i'\right|^2, \overline{v}_i' = \overline{v}_i - \overline{V}$$

2.5   VAN DER WAALS EQUATION (CORRESPONDING STATES LAW)

In 1873, Van der Waals proposed an equation of state is qualitatively good description of liquid and gaseous systems. It is for one mole (one mole) is:

$$\left(p + \frac{a}{v^2}\right)(v - b) = RT \tag{2.2}$$

Note that at $p > \frac{a}{v^2}$ and $v > b$ this equation becomes the equation of state of ideal gas

$$pv = RT \tag{2.3}$$

Van der Waals equation can be obtained from the Clapeyron equation of Mendeleev by an amendment to the magnitude of the pressure a/v^2 and the amendment b to the volume, both constant a and b independent of T and v but dependent on the nature of the gas.

The amendment b takes into account:

1) the volume occupied by the molecules of real gas (in an ideal gas molecules are taken as material points, not occupying any volume);

2) so-called "dead space," in which cannot penetrate the molecules of real gas during motion, that is, volume of gaps between the molecules in their dense packing.

Thus, $b = v_{i\bar{i}e.} + v_{\varsigma\grave{a}\varsigma}$ (Fig. 2.3). The amendment a/v^2 takes into account the interaction force between the molecules of real gases. It is the internal pressure, which is determined from the following simple considerations. Two adjacent elements of the gas will react with a force proportional to the product of the quantities of substances enclosed in these elementary volumes.

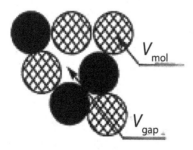

FIGURE 2.3   Location scheme ofmoleculesin a real gas.

TABLE 2.1   The Molecular Weight ($[\mu] = \kappa c / \kappa моль$) of Some Gases

gas	N	Ar	H_2	O_2	CO	CO_2	ammonia	air
m	28	40	2	32	28	44	17	29

Therefore, the internal pressure $p_{вн}$ is proportional to the square of the concentration n:

$$p_{вн} \sim n^2 \sim \rho^2 \sim \frac{1}{v^2},$$

where, r–the gas density.

Thus, the total pressure consists of internal and external pressures:

$$p + p_{эн} = p + \frac{a}{v^2}$$

Equation (2.3) is the most common for an ideal gas.

Under normal physical conditions ($p_0 = 0,1013 МПа, t_0 = 0^0 C$) $v\mu = 22,4 м^3 /(кмоль \cdot K^0)$, and then from Eq. (2.3) we obtain:

$$R\mu = \frac{pv\mu}{T} = \frac{0,1013 \cdot 10^6 \cdot 22,4}{273} = 8314 \frac{\partial \ni C}{KMOль \cdot K^0}$$

Knowing $R\mu$ we can find the gas constant for any gas with the help of the value of its molecular mass μ (Table 2.1):

$$R = \frac{R\mu}{\mu} = \frac{8314}{\mu}$$

For gas mixture with mass M state equation has the form:

$$pv = MR_{CM}T = \frac{8314MT}{ì_{CM}} \tag{2.4}$$

where $R_{cм}$–gas constant of the mixture.

The gas mixture can be given by the mass proportions g_i, voluminous r_i or mole fractions n_i, respectively, which are defined as the ratio of mass m_i, volume v_i or number of moles N_i of i gas to total mass M, volume v or number of moles N of gas mixture. Mass fraction of component is $g_i = \frac{m_i}{M}$, where $i = 1, n$. It is obvious that $M = \sum\limits_{i=1}^{n} m_i$ and $\sum\limits_{i=1}^{n} g_i = 1$. The volume fraction is $r_i = \dfrac{v_i}{v_{CM}}$, , where

v_i –partial volume of component mixtures.

Similarly, we have $\sum\limits_{i=1}^{n} v_i = v_{CM}, \sum\limits_{i=1}^{n} r_i = 1$.

Depending on specificity of tasks the gas constant of the mixture may be determined as follows:

$$R_{CM} = \sum_{i=1}^{n} g_i R_i \quad ; \qquad R_{CM} = \dfrac{1}{\sum\limits_{i=1}^{n} r_i R_i^{-1}}$$

If we know the gas constant R_{CM}, the seeming molecular weight of the mixture

is equal to:

$$\mu_{CM} = \dfrac{8314}{R_{CM}} = \dfrac{8314}{\sum\limits_{i=1}^{n} g_i R_i} = 8314 \sum_{i=1}^{n} r_i R_i^{-1}$$

The pressure of the gas mixture p is equal to the sum of the partial pressures of individual components in the mixture p_i:

$$p = \sum_{i=1}^{i=1n} p_i \tag{2.5}$$

Partial pressure p_i–pressure that has gas, if it is one at the same temperature fills the whole volume of the mixture ($p_i v_{CM} = RT$)..

With various methods of setting the gas mixture partial pressures

$$p_i = pr_1; p_1 = \dfrac{pg_i \mu_{cm}}{\mu_i} \; p_i = pr_i \; ; \tag{2.6}$$

From the Eq. (2.6) we see that for the calculation of the partial pressures p_i necessary to know the pressure of the gas mixture, the volume or mass fraction i of the gas component, as well as the molecular weight of the gas mixture m and the molecular weight of i of gas μ_i.

The relationship between mass and volume fractions are written as follows:

$$g_i = \dfrac{m_i}{m_{CM}} = \dfrac{\rho_i v_i}{\rho_{CM} v_{CM}} = \dfrac{R_{CM}}{R_i} r_i = \dfrac{\mu_i}{\mu_{CM}} r_i$$

We rewrite Eq. (2.2) as

$$v^3 - \left(b + \frac{RT}{p}\right)v^2 + \frac{a}{p}v - \frac{ab}{p} = 0 \, .$$

(2.7)

When $p = p_k$ and $T = T_k$, where p_k and T_k –critical pressure and temperature, all three roots of Eq. (2.7) are equal to the critical volume v_k

$$v^3 - \left(b + \frac{RT_k}{p_k}\right)v^2 + \frac{a}{p_k}v - \frac{ab}{p_k} = 0 \, .$$

(2.8)

Because the $v_1 = v_2 = v_3 = v_k$, then Eq. (2.8) must be identical to the equation

$$\left(v - v_1\right)\left(v - v_2\right)\left(v - v_3\right) = \left(v - v_k\right)^3 = v^3 - 3v^2v_k + 3vv_k^2 - v_k^3 = 0 \, .$$

(2.9)

Comparing the coefficients at the equal powers of v in both equations leads to the equalities

$$b + \frac{RT_k}{p_k} = 3v_k; \frac{a}{p_k} = 3v_k^2; \frac{ab}{p_k} = v_k^3 \, .$$

(2.10)

Hence

$$a = 3v_k^2 p_k; b = \frac{v_k}{3}$$

(2.11)

Considering (the Eq. (2.10)) as equations for the unknowns p_k, v_k, T_k, we obtain

$$p_k = \frac{a}{27b^2}; v_k = 3b; T_k = \frac{8a}{27bR} \, .$$

(2.12)

From Eqs. (2.10) and (2.11) or (2.12) we can find the relation

$$\frac{RT_k}{p_k v_k} = \frac{8}{3}$$

(2.13)

Instead of the variables p, v, T let's introduce the relationship of these variables to their critical values (leaden dimensionless parameters).

$$\pi = \frac{p}{p_k}; \, \omega = \frac{v}{V_k}; \, \tau = \frac{T}{T_k} \, .$$

(2.14)

Substituting Eqs. (2.12) and (2.14) in (2.7) and using Eq. (2.13), we obtain

$$\left(\pi p_k + \frac{3v_k^2 p_k}{\omega^2 v_k^2} \right)\left(\omega v_k - \frac{v_k}{3} \right) = RT_k \tau$$

$$\left(\pi + \frac{3}{\omega^2} \right)(3\omega - 1) = 3\frac{RT_k}{p_k v_k}\tau$$

$$\left(\pi + \frac{3}{\omega^2} \right)(3\omega - 1) = 8\tau, \qquad\qquad (2.15)$$

In Eq. (2.15), a and b are not permanent, depending on the nature of the gas. That is, if the units of measurement of pressure, volume and temperature are used as their critical values (use the leaden parameters), the equation of state is the same for all substances.

This condition is called the *law of corresponding states.*

KEYWORDS

- **Capillary Effects**
- **Carbon Nanotubes**
- **Continuum Hypothesis**
- **Microscale Systems**
- **Nanoscale Systems**

CHAPTER 3

RHEOLOGICAL PROPERTIES AND STRUCTURE OF THE LIQUID IN NANOTUBES

CONTENTS

3.1   SLIPPAGE OF THE FLUID PARTICLES NEAR THE WALL

According to the Navier boundary condition, the velocity slip is proportional to fluid velocity gradient at the wall:

$$v\big|_{y=0} = L_S dv / dy\big|_{y=0} \qquad (3.1)$$

Here and in Fig. 3.1, L_s represents the "slip length" and has a dimension of length.

Because of the slippage, the average velocity in the channel $\langle v_{pdf} \rangle$ increases.

In a rectangular channel (of width >> height h and viscosity of the fluid η) due to an applied pressure gradient of dp/dex:

$$\langle v_{pdf} \rangle = \frac{h^2}{12\varsigma}\left(-\frac{dp}{dx}\right)\left(1+\frac{6L_S}{h}\right) \qquad (3.2)$$

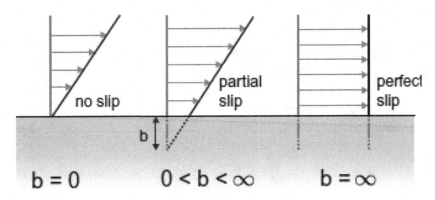

FIGURE 3.1   Three cases of slip flow (with slip length b).

The results of molecular dynamics simulation for nanosystems with liquid, show that a large slippage lengths (of the order of microns) should occur in the carbon nanotubes of nanometer diameter and, consequently, can increase the flow rate by three orders of ($6L_S / h > 1000$). Thus, the flow with slippage is becoming more and more important for hydrodynamic systems of small size.

The results of molecular dynamics simulation of unsteady flow of mixtures of water–water vapor, water and nitrogen in a carbon nanotube are reported previously. Based on these studies, a flow of water through carbon nanotubes with different diameters at temperature of 300°K was considered.

Carbon nanotubes have been considered "zigzag" with chiral vectors (20, 20), (30, 30) and (40, 40), corresponding to pipe diameters or 2712, 4068 and 5424 nm, respectively.

The value of the flow rate and the system pressure, which varies in the range of 600–800 bars, were high enough to ensure complete filling of the tubes. This pressure was achieved by the total number of water molecules 736, 904, and 1694, respectively.

The effects of slippage of various liquids on the surface of the nanotube were studied in detail.

The length of slip, calculated using the current flow velocity profiles of liquid, shown in Fig. 3.2, were 11, 13, and 15 nm for the pipes of 2712, 4068 and 5424 nm, respectively. The dotted line marked by theoretical modeling data. The vertical lines indicate the position of the surface of carbon nanotubes.

It was found out that as the diameter decreases, the speed of slippage of particles on the wall of nanotube also decreases. The report attributes this to the increase of the surface friction.

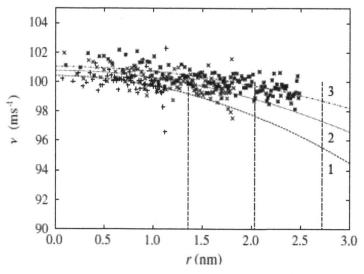

FIGURE 3.2   Time average streaming velocity profiles of water in a carbon nanotubes of different diameter: 2.712 nm, curve 1; 4.068 nm, curve 2; 5.424 nm, curve 3.

Experiments with various pressure drops in nanotubes demonstrated slippage of fluid in micro and nanosystems. The most remarkable were the two recent experiments, which were conducted to improve the flow characteristics of carbon nanotubes with the diameters of 2 and 7 nm, respectively.

In the membranes in which the carbon nanotubes were arranged in parallel, there was a slip of the liquid in the micrometer range. This led to a significant increase in flow rate, up to three and four orders of magnitude.

In the experiments for the water moving in microchannels on smooth hydro-phobic surfaces, there are sliding at about 20 nm. If the wall of the channel is not smooth but twisty or rough (and hydrophobic) such a structure would lead to an accumulation of air in the cavities and become super hydrophobic (with contact angle greater than 160°). It is believed that this leads to creation of contiguous areas with high and low slippage, which can be described as "effective slip length." This effective length of the slip occurring on the rough surface can be several tens of microns, which was confirmed experimentally.

It should be noted that for practical use of advantages of nanotubes with slip-page is necessary to solve many more problems. Scientists have already shown that hydrophobic surfaces tend to form bubbles. On the other hand, the surfaces used by most researchers were rough, but the use of smooth surfaces could generally reduce the formation of bubbles.

Another possible problem is filling of the hydrophobic systems with liquid. Filling of micron size hydrophobic capillaries is not a big problem, because pressure of less than 1 atm is sufficient. Capillary pressure, however, is inversely proportional to the diameter of the channel, and filling for nanochannels can be very difficult.

3.2   THE DENSITY OF THE LIQUID LAYER NEAR A WALL OF CARBON NANOTUBE

Scientists showed radial density profiles of oxygen averaged in time and hydro-gen atoms in the "zigzag" carbon nanotube with chiral vector (20, 20) and a radius $R = 1,356$ nm (Fig. 3.3). The distribution of molecules in the area near the wall of the carbon nanotube indicated a high density layer near the wall of the carbon nanotube. Such a pattern indicates the presence of structural heterogeneity of the liquid in the flow of the nanotube.

In Fig. 3.3, $\rho^*(r) = \rho(r)/\rho(0)$, whereas 2712 nm diameter pipe is completely filled with water molecules at 300°K. The overall density is $\rho(0) = 1000 \kappa z / \text{м}^3$. The arrows denote the location of distinguishable layers of the water molecules and the vertical line is the position of the CNT wall.

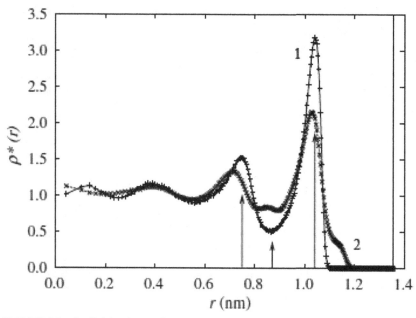

FIGURE 3.3   Radial density profiles of oxygen (curve 1) and hydrogen (curve 2) atoms.

The distribution of molecules in the region of $0.95 \leq r \leq R$ nm indicates a high density layer near the wall carbon nanotube. Such a pattern indicates the presence of structural heterogeneity of the liquid in the flow of the nanotube.

Figure 3.4 shows the scheme of the initial structure and movement of water molecules in:

(a) the model carbon nanotube;

(b) the radial density profile of water molecules inside nanotubes with different radii and;

(c) the velocity profile of water molecules in a nanotube chirality (60,60).

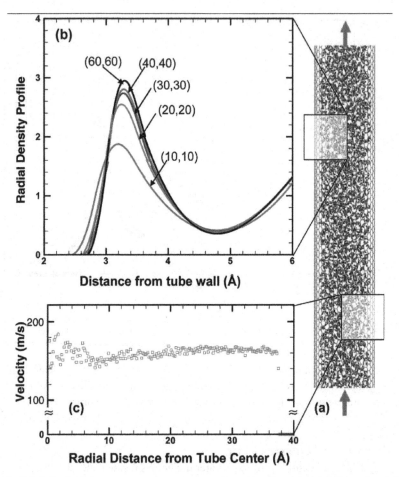

FIGURE 3.4   Initial structure and movement of water molecules: (a) Schematic of the initial structure and transport of water molecules in a model CNT. (b) The radial density profile (RDP) of water molecules inside CNTs with different radii. (c) The representative radial velocity profile (RVP) of water molecules inside a (60,60) nanotube.

It should be noted that the available area for molecules of the liquid is less than the area bounded by a solid wall, primarily due to the van der Waals interactions.

3.3   THE EFFECTIVE VISCOSITY OF THE LIQUID IN A NANOTUBE

There is a significant increase in the effective viscosity of the fluid in the nano volumes compared to its macroscopic value. However, the effective viscosity of the liquid in a nanotube depends on the diameter of the nanotube.

The effective viscosity of the liquid in a nanotube is defined as follows.

A conformity nanotubes filled with liquid, containing crystallites with the same size tube filled with liquid, can be considered as a homogeneous medium (i.e., without considering the crystallite structure), in which the same pressure drop and flow rate of Poiseuille flow is realized. The viscosity of a homogeneous fluid, whichensures the coincidence of these parameters, called the effective viscosity of the flow in the nanotube.

While flowing in the narrow channels of width less than 2 nm, water behaves like a viscous liquid. In the vertical direction, water behaves as a rigid body, and in a horizontal direction it maintains its fluidity.

It is known that at large distances the van der Waals interaction has a magnetic tendency and occurs between any molecules like polar as well as non-polar. At small distances it is compensated by repulsion of electron shells. Van der Waals interaction decreases rapidly with distance. Mutual convergence of the particles under the influence magnetic forces continues until these forces are balanced with the increasing forces of repulsion.

Knowing the deceleration of the flow (Fig. 3.4) of water a, the effective shear stress between the wall of the pipe length l and water molecules can be calculated by;

$$\tau = Nma \,/\left(2\pi Rl\right)$$
(3.3)

Here, the shear stress is a function of tube radius and flow velocity $\bar{v}$, and m is mass of water molecules. The average speed is related to volumetric flow $\bar{v} = Q/\left(\pi R^2\right)$.

Denoting n_0 the density of water molecules number, we can calculate the shear stress in the form of:

$$\tau\big|_{r=R} = n_0 mRa/2$$
(3.4)

Figure 3.5 shows the results of calculations concerning the influence of the size of the tube R_0 on the effective viscosity (squares) and shear stress τ (triangles), when the flow rate is approximately 165 m/sec.

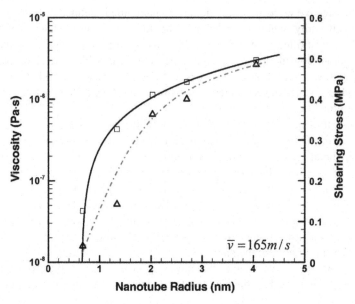

FIGURE 3.5   Size effect of shearing stress (triangle) and viscosity (square).

According to classical mechanics of liquid flow at different pressure drops Δp along the tube length l is given by Poiseuille formula;

$$Q_P = \frac{\pi R^4 \Delta p}{8\eta l},$$ (3.5)

Therefore,

$$\tau = \frac{\Delta p R}{2l}$$ (3.6)

and the effective viscosity of the fluid can be estimated as $\eta = \tau \cdot R / (4\bar{v})$.

The change in the value of shear stress directly causes the dependence of the effective viscosity of the fluid from the pipe size and flow rate. In this case the effective viscosity of the transported fluid can be determined from Eqs. (3.5) and (3.6) as

$$\eta = \frac{\tau \cdot R}{4\bar{v}}$$ (3.7)

It should be noted that the magnitude of the shear stress τ is relatively small in the range of pipe sizes considered. This indicates that the surface of carbon nanotubes is very smooth and the water molecules can easily slide through it.

In fact, shear stress is primarily due to van der Waals interaction between the solid wall and the water molecules. It is noted that the characteristic distance between the near-wall layer of fluid and pipe wall depends on the equilibrium distance between atoms O and C and the distribution of the atoms of the solid wall and bend of the pipe.

From Fig. 3.5, the effective viscosity η increases by two orders of magnitude when R_0 changing from 0.67 to 5.4 nm.

According to Eqs. (3.5)–(3.7), the effective viscosity can be calculated as $\eta = \dfrac{\pi R^4 \Delta p}{8Q}$. The results of calculations are shown in Fig. 3.6.

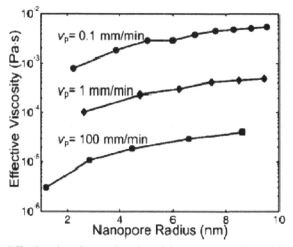

FIGURE 3.6   Effective viscosity as a function of the nanopore radius and the loading rate.

The dependence of the shear stress on the flow rate is illustrated in Fig. 3.7. For the tube (20,20) τ increases with $\bar{v}$. The growth rate slowed down at higher values $\bar{v}$.

At high speeds $\bar{v}$, while water molecules are moving along the surface of the pipe, the liquid molecules do not have enough time to fully adjust their positions to minimize the free energy of the system. Therefore, the distance between adjacent carbon atoms and water molecules may be less than the equilibrium van der Waals distances. This leads to an increase in van der Waals forces of repulsion and leads to higher shear stress.

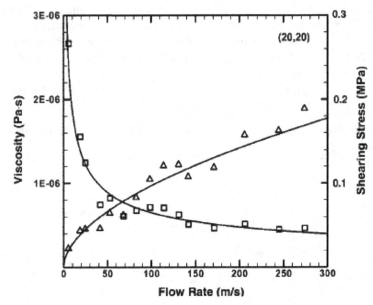

FIGURE 3.7   Flow rate effect of shearing stress (triangle) and viscosity (square).

It should be noted that even though the equation for viscosity is based on the theory of the continuum, it can be extended to a complex flow to determine the effective viscosity of the nanotube.

Figure 3.7 shows dependence of h on $\bar{v}$ inside of the nanotube (20, 20). It is shown that h decreases sharply with increasing flow rate and begins to approach a definite value when $\bar{v} > 150$ m/sec. For the current pipe size and flow rate ranges of $\eta \sim 1/\sqrt{v}$, this trend is because of $\tau - \bar{v}$, contained in Fig. 3.7. According to Fig. 3.6, high-speed effects are negligible.

One can easily see that the dependence of viscosity on the size and speed is consistent qualitatively with the results of molecular dynamic simulations. In all studied cases, the viscosity is much smaller than its macroscopic analogy. As the radius of the pores varies from about 1 nm to 10 nm, then the value of the effective viscosity increases by an order of magnitude. A more significant change occurs when increasing the speed of 0.1 mm/min up to 100 mm/min. This results in a change in the value of viscosity h, respectively, by 3–4 orders. The discrepancy between simulation and test data can be associated with differences in the structure of the nanopores and liquid phase.

Figure 3.8 shows the viscosity dependence of water (calculated by the method of DM, the diameter of the CNT). The viscosity of water, as shown in the Fig. 8, increases monotonically with increasing diameter of the CNT.

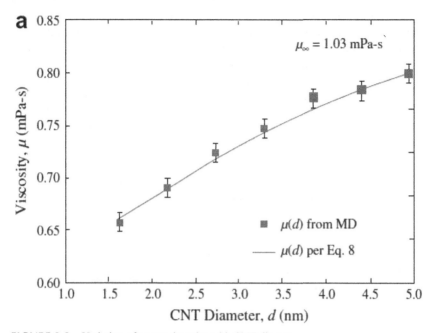

FIGURE 3.8   Variation of water viscosity with CNT diameter.

3.4   THE RELEASE OF ENERGY DUE TO THE COLLAPSE OF THE NANOTUBE

Scientists theoretically predicted the existence of a "domino effect" in single-walled carbon nanotube.

Squashing can occur at one end by two rigid movements of narrow graphene planes (about 0.8 nm in width and 8.5 nm in length). This can rapidly (at a rate exceeding 1 km/s) release its stored energy by collapsing along its length like a row of dominoes. The effect resembles a tube of toothpaste squeezing itself (Fig. 3.9).

The structure of a single-walled carbon nanotube has two possible stable states: circular or collapsed. Scientists realized that for nanotubes wider than 3.5 nanometers, the circular state stores more potential energy than the collapsed state as a result of van der Waal's forces. He performed molecular dynamics simulations to find out what would happen if one end of a nanotube was rapidly collapsed by clamping it between two graphene bars.

This phenomenon occurs with the release of energy, and thus allows for the first time to talk about carbon nanotubes as energy sources. This effect can also be used as an accelerator of molecules.

The tube collapses at the same time not over its entire length, and sequentially, one after the other carbon ring, starting from the end, which is tightened (Fig. 3.9). It happens just like a domino collapses, arranged in a row (this is known as the "domino effect"). The role of bone dominoes performing here as ring of carbon atoms forming the nanotube, and the nature of this phenomenon is quite different. Recent studies show that for nanotubes with diameters ranging from 2 to 6 nm, there are two stable equilibrium states;

• Cylindrical (tube no collapses) and;
• Compressed (imploded tube) with different values of potential energy (the difference between which and can be used as an energy source).

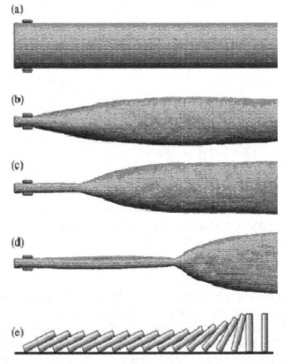

FIGURE 3.9   "Domino effect" in a carbonnanotube. (a) The initial form of carbon nanotubes is cylindrical. (b) One end ofthe tubeis squeezed. (c), (d) Propagation of domino waves; the configuration of the nanotube 15 and 25 picoseconds after the compression of its end. (e) Schematic illustration of the "domino effect" under the influence of gravity.

The switching between these two states with the subsequent release of energy occurs in the form of waves arising domino effect. The scientists have shown that such switching is carried out in carbon nanotubes with diameters of 2 nm.

A theoretical study of the "domino effect" was conducted by using a special method of classical molecular dynamics, in which the interaction between carbon atoms was described by van der Waals forces.

The main reason for the observed effect, is the potential energy of the van der Waals interactions (which "collapses" the nanotube) and the energy of elastic deformation (which seeks to preserve the geometry of carbon atoms). This eventually leads to a bistable (collapsed and no collapsed) configuration carbon nanotube.

Thus, "domino effect" wave can be produced in a carbon nanotube with a relatively large diameter (more than 3.5 nm), because only in such a system the potential energy of collapsing structures may be less than the potential energy of the "normal" nanotube. In other words, the cylindrical structure and collapsing nanotubes with large diameters are, respectively, of the meta-stable and stable states.

The potential energy of a carbon nanotube with a propagating wave at the "domino effect" represented in Fig. 3.10a. This Fig. 3.10 shows three sections:

- The first (from 0 ps to 10 ps) are composed of elastic strain energy, whichappears due to changes in the curvature of the walls of the nanotubes in the process of collapsing.
- The second region (from 10 ps to 35 ps) corresponds to the "domino effect."
- Finally, the third segment (from 35 ps to 45 ps) corresponds to the process is ended "domino."

The spread "domino effect" waves are a process that goes with the release of energy (about 0.01 eV per atom of carbon). This is certainly not comparable in any way with the degree of energy yield in nuclear reactions, but the fact of power generation carbon nanotube is obvious.

Calculations show that the wave of dominoes in a tube with diameter of 4–5 nm is about 1 km/s (as seen from the Fig. 3.10b) and non-linear manner depending on its geometry (the diameter and chirality). The maximum effect should be observed in the tube with a diameter slightly less than 4.5 nm (considering carbon rings to collapse at a speed of 1.28 km/sec). The theoretical dependence shows the blue solid line. And now an example of how energy is released in such a system with a "domino effect" can be used in nano-devices. Scientists offer an original way to use sort of "nanogun" (Fig. 3.10a). Imagine that at our disposal is a carbon nanotube with chirality (55.0) and related to the observation of the dominoes in diameter. Put inside a nanotube fullerene C_{60}. With a little imagination, this can be considered a carbon nanotube as the gun trunk, and the molecule (as its shell).

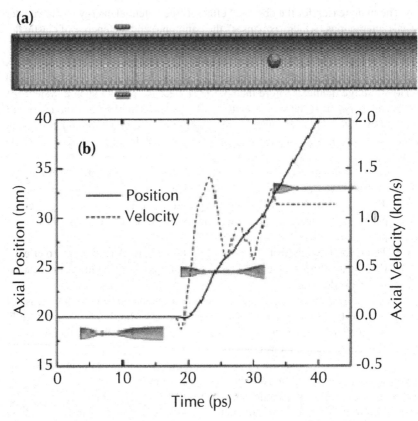

FIGURE 3.10 Nano cannon scheme acting on the basis of "domino effect" in the incision. (a)Inside a carbon nanotube (55.0) is the fullerene molecule C_{60}. (b) The initial position and velocity of the departure of the "core" (a fullerene molecule), depending on the time. The highest rate of emission of C_{60} (1.13 km /s) comparable to the velocity of the domino wave.

KEYWORDS

- **Fluid particles**
- **Liquid**
- **Naotubes**
- **Rheological properties**
- **Slippage**
- **Viscosity**

CHAPTER 4

FLUID FLOW IN NANOTUBES

CONTENTS

4.1 INTRODUCTION

Scientists from the University of Wisconsin-Madison (USA) managed to prove that the laws of friction for the nanostructures do not differ from the classical laws.

The friction of surface against the surface in the absence of the interlayer between the liquid materials (so-called dry friction) is created by irregularities in the given surfaces that rub one another, as well as the interaction forces between the particles that make up the surface.

As part of their study, the researchers built a computer model that calculates the friction force between nano-surfaces (Fig. 4.1). In the model, these surfaces were presented simply as a set of molecules for which forces of intermolecular interactions were calculated.

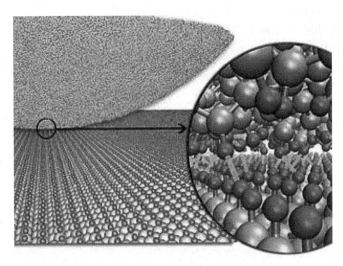

FIGURE 4.1   Computer model of friction at the nanoscale. The right shows the surfaces of interacting particles (Friction laws at the nanoscale).

As a result, scientists were able to establish that the friction force is directly proportional to the number of interacting particles. The researchers propose to consider this quantity by analog of so-called true macroscopic contact area. It is known that the friction force is directly proportional to this area (it should not be confused with common area of the contact surfaces of the bodies).

In addition, the researchers were able to show that the friction surface of the nano-surfaces can be considered within the framework of the classical theories of friction of non-smooth surfaces.

A literature review shows that nowadays, molecular dynamics and mechanics of the continuum in are the main methods of research of fluid flow in nanotubes.

Although the method of molecular dynamics simulations is effective, it at the same time requires enormous computing time especially for large systems. Therefore, simulation of large systems is more reasonable to carry out nowadays by the method of continuum mechanics [52, 53, 76, 77, 81–83].

In the work of Morten Bo Lindholm Mikkelsen et al. [48], the fluid flow in the channel is considered in the framework of the continuum hypothesis. The Navier-Stokes equation was used and the velocity profile was determined for Poiseuille flow.

In the work of Thomas John and McGaughey Alan [73], the water flow by means of pressure differential through the carbon nanotubes with diameters ranging from 1.66 to 4.99 nm is researched using molecular dynamics simulation study. For each nanotube the value enhancement predicted by the theory of liquid flow in the carbon nanotubes is calculated. This formula is defined as a ratio of the observed flow in the experiments to the theoretical values without considering slippage on the model of Hagen-Poiseuille. The calculations showed that the enhancement decreases with increasing diameter of the nanotube.

Important conclusion of the Thomas John and McGaughey Alan [73] is that by constructing a functional dependence of the viscosity of the water and length of the slippage on the diameter of carbon nanotubes, the experimental results in the context of continuum fluid mechanics can easily be described. The aforementioned is true even for carbon nanotubes with diameters of less than 1.66 nm.

The theoretical calculations use the following formula for the steady velocity profile of the viscosity η of the fluid particles in the CNT under pressure gradient $\partial p / \partial z$:

$$v(r) = \frac{R^2}{4\varsigma}\left[1 - \frac{r^2}{R^2} + \frac{2L_S}{R}\right]\frac{\partial p}{\partial z} \tag{4.1}$$

The length of the slip, which expresses the speed heterogeneity at the boundary of the solid wall and fluid is defined as by Joseph [33] and Sokhan [67]:

$$L_S = \frac{v(r)}{dv / dr}\bigg|_{r=R} \tag{4.2}$$

Then the volumetric flow rate, taking into account the slip Q_S is defined as:

$$Q_S = \int_0^R 2\eth r \cdot v(r)\,dr = \frac{\eth(d/2)^4 + 4(d/2)^3 . L_S}{8\varsigma} \cdot \frac{\partial p}{\partial z} \tag{4.3}$$

Equation (4.3) is a modified Hagen-Poiseuille equation, taking into account slippage. In the absence of slip $L_s = 0$ (the Eq. (4.3)) coincides with the Hagen-Poiseuille flow (the Eq. (3.5)) for the volumetric flow rate without slip

Q_P. In the works of many scientists the parameter enhancement flow $\mathcal{E}$ is introduced. It is defined as the ratio of the calculated volumetric flow rate of slippage to Q_P (calculated using the effective viscosity and the diameter of the CNT). If the measured flux is modeled using the Eq. (4.3), the degree of enhancement takes the form:

$$\mathcal{E} = \frac{Q_S}{Q_P} = \left[1 + 8\frac{L_S(d)}{d}\right]\frac{\eta_\infty}{\eta(d)} \tag{4.4}$$

where, $d = 2R$–diameter of CNT; η_∞ – viscosity of water; $L_S(d)$– CNT slip length depending on the diameter; $\eta(d)$– the viscosity of water inside CNTs depending on the diameter.

–   If $\eta(d)$ finds to be equal to η_∞, then the influence of the effect of slip on $\mathcal{E}$ is significant,

–   If $L_S(d) \geq d$. If $L_S(d) < d$ and $\eta(d) = \eta_\infty$, then there will be no significant difference compared to the Hagen-Poiseuille flow with no slip.

Table 4.1 shows the experimentally measured values of the enhancement water flow. Enhancement flow factor and the length of the slip were calculated using the equations given above.

Figure 4.2 depicts the change in viscosity of the water and the length of the slip in diameter. As can be seen from the figure, the dependence of slip length to the diameter of the nanotube is well described by the empirical relation.

TABLE 4.1   Experimentally measured values of the enhancement water flow

Nanosystems	Diameter (nm)	Enhancement, $\mathcal{E}$	Slip length, L_S, (nm)
carbonnanotubes	300–500	1	0
	44	22–34	113–177
carbonnanotubes	7	10^4–10^5	3900–6800
	1.6	560–9600	140–1400

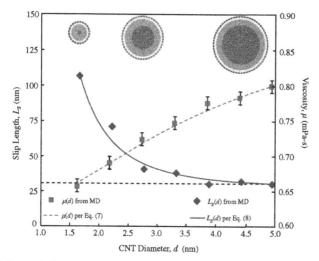

FIGURE 4.2   Variation of viscosity and slip length with CNT diameter.

$$L_S(d) = L_{S,\infty} + \frac{C}{d^3}$$

$$(4.5)$$

where, $L_{S,\infty}$ =30 nm – slip length on a planesheet ofgraphene; C – const.

Figure 4.3 shows dependence of the enhancementof the flow rate $\mathcal{E}$ on the diameter for all seven CNTs.

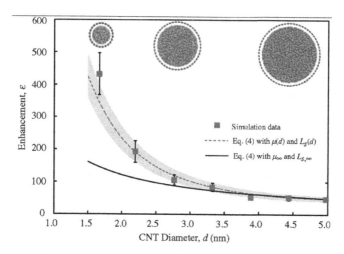

FIGURE 4.3   Flow enhancement as predicted from MD simulations [73].

There are three importantfeatures inthe results:
(1) the enhancement of the flow decreases with increasing diameter of the CNT.
(2) with increasing diameter, the value tends to the theoretical value of Eqs. (4.4) and (4.5) with a slip $L_{s\infty} = 30$ nm and the effective viscosity $\mu(d) = \mu_\infty$. The dotted line showed the curve of 15% in the second error in the theoretical data of viscosity and slip length.
(3) the change ε in diameter of CNTs cannot be explained only by the slip length.

To determine the dependence the volumetric flow of water from the pressure gradient along the axis of single-walled nanotube with the radii of 1.66, 2.22, 2.77, 3.33, 3.88, 4.44 and 4.99 nm in Thomas John and McGaughey Alan [73],the method of molecular modeling was used. Snapshot of the water-CNT is shown in Fig. 4.4.

Figure 4.4 shows the results of calculations to determine the pressure gradient along the axis of the nanotube with the diameter of 2.77 nm and a length of 20 nm. Change of the density of the liquid in the cross sections was less than 1%.

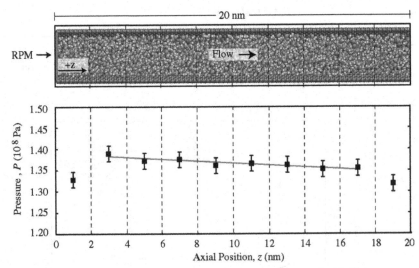

FIGURE 4.4   Axial pressure gradient inside the 2.77 nm diameter CNT [73].

Figure 4.5 shows the dependence of the volumetric flow rate from the pressure gradient for all seven CNTs. The flow rate ranged from 3–14 m/sec. In the range considered here the pressure gradient $(0 - 3).10^9$ *atm/m* Q (*pl/sek* $= 10^{-15}$ *m³/sek*) is directly proportional to $\partial p / \partial z$. Coordinates of chirality for each CNT are indicated in the figure legend. The linearity of the relations between flow and pressure gradient confirms the validity of calculations of the Eq. (4.3).

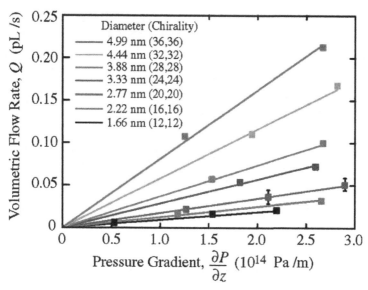

FIGURE 4.5 Relationship between volumetric liquid flow rate in carbon nanotubes with different diameters and applied pressure gradient

Figure. 4.6 shows the profile of the radial velocity of water particles in the CNT with diameter 2.77 nm. The vertical dotted line at 1.38 nm marked surface of the CNT. It is seen that the velocity profile is close to a parabolic shape.

In contrast to previous work [74], the flow of water under a pressure gradient considered in the single-walled nanotubes of "chair" type of smaller radii: 0.83, 0.97, 1.10, 1.25, 1.39 and 1.66 nm.

Figure 4.7 shows the dependence of the mean flow velocity $\bar{v}$ on the applied pressure gradient $\Delta P / L$ in the long nanotubes is equal to 75 nm at 298 K. A similar picture pattern occurs in the tube with the length of 150 nm.

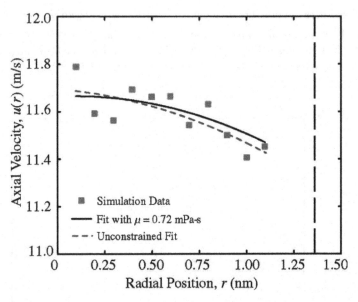

FIGURE 4.6   Radial velocity profile inside 2.77 nm diameter CNT.

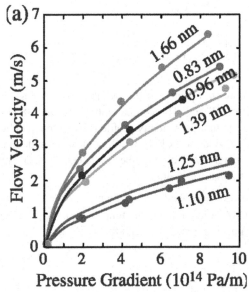

FIGURE 4.7   Relationship between average flow velocity and applied pressure gradient for the 75 nm long CNTs.

As one can see, there is conformance with the Darcy law, the average flow rate for each CNT increases with increasing pressure gradient. For a fixed value of $\Delta p / L$, however, the average flow rate does not increase monotonically with increasing diameter of the CNTs, as follows from Poiseuille equation. Instead, when at the same pressure gradient, decrease of the average speed in a CNT with the radius of 0.83 nm to a CNT with the radius of 1.10 nm, similar to the CNTs 1.10 and 1.25 nm, then increases from a CNT with the radius of 1.25 nm to a CNT of 1.66 nm.

The non-linearity between $\bar{v}$ and $\Delta P / L$ are the result of inertia losses (i.e., insignificant losses) in the two boundaries of the CNT. Inertial losses depend on the speed and are caused by a sudden expansion, abbreviations, and other obstructions in the flow.

Molecular modeling of many researchers shows that the Eq. (4.1) (Poiseuille parabola) describes the velocity profile of liquid in a nanotube when the diameter of a flow is 5–10 times more than the diameter of the molecule (≈ 0.17 nm for water).

In Fig. 4.8, the effect of slip on the velocity profile at the boundary of radius R of the pipe and fluid is shown. When $L_s = 0$ the fluid velocity at the wall vanishes, and the maximum speed (on the tube axis) exceeds flow speed twice.

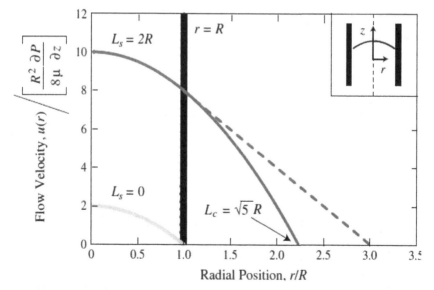

FIGURE 4.8 No-slip Poiseuille flow and slip Poiseuille flow through a tube.

The Fig. 4.8 shows the velocity profiles for Poiseuille flow without slip (L_s = 0) and with slippage L_s = 2R. The flow rate is normalized to the speed corresponding to the flow without slip. Thick vertical lines indicate the location of the pipe wall. The thick vertical lines indicate the location of the tube wall.

Velocity of the liquid on a solid surface can also be quantified by the coefficient of slipL_c. The coefficient of slippage is a difference between the radial positions in which the velocity profile would be zero. Slip coefficient is equal to

$$L_C = \sqrt{R^2 + 2R}_s = \sqrt{5}R.$$

For linear velocity profiles (e.g., Couette flow), the length of the slip and slip rate are equal. These values are different for the Poiseuille flow.

Figure 4.9 shows dependence of the volumetric flow rate Q from the pressure gradient $\partial p / \partial z$ in long nanotubes with diameters between 1.66 nm and 6.93 nm. Pressure gradient Q is proportional to $\partial p / \partial z$. As in the Poiseuille flow, volumetric flow rate increases monotonically with the diameter of CNT at a fixed pressure gradient. Magnitudes of calculations error for all the dependencies are similar to the error for the CNT diameter 4.44 nm (marked in the Fig. 4.9).

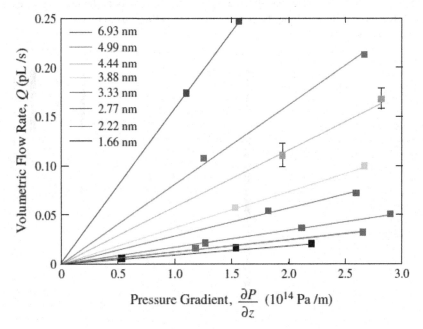

FIGURE 4.9   Volumetric flow rate in CNTs versus pressure gradient.

Researchers [79] considered steady flow of incompressible fluids in a channel width $2h$ under action of the force of gravity ρg or pressure gradient $\partial p / \partial y$ (which is described by the Navier-Stokes equations). The velocity profile has a parabolic form:

$$U_y(z) = \frac{\rho g}{2\eta} \cdot \left[(\delta + h)^2 - z^2 \right]$$

where, δ – length of the slip, which is equal to the distance from the wall to the point at which the velocity extrapolates to zero.

4.2 MODELING IN NANOHYDROMECANICS

We take into consideration that the mean free path of gas under normal conditions is 65 нм, and the distance between the particles – $3,3$ нм ::

$$\frac{4}{3}\pi\delta^3 n = 1, \qquad \delta_{станд} = 3,3 \text{ нм}, \qquad \lambda = \frac{1}{\sqrt{2}\pi d^2 n}, \qquad \lambda_{станд} = 65 \text{ нм},$$

where, n – the concentration of moleculesin the air, Knudsen number $= \lambda/L$; L – the characteristic size; d – diametermicrotubes.

Let's consider the fluid flow through the nanotube. Molecules of a substance in a liquid state are very close to each other (Fig. 4.10).

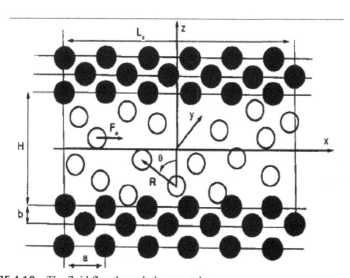

FIGURE 4.10   The fluid flowthrough the nanotube.

Most liquid's molecules have a diameter of about 0.1 nm. Each molecule of the fluid is "squeezed" on all sides by neighboring molecules and for a period of time (10^{-10}–10^{-13}) fluctuates around certain equilibrium position, which itself from time to time is shifted in distance commensuration with the size of molecules or the average distance between molecules l_{cp}:

$$l_{cp} \approx \sqrt[3]{\frac{1}{n_0}} = \sqrt[3]{\frac{\mu}{N_A \rho}},$$

where, n_0 – Numberof molecules per unitvolume of fluid; N_A–Avogadro's number; ρ –fluid density; μ –molar mass.

Estimates show that one cubic of nano water contains about 50 molecules. This gives a basis to describe the mass transfer of liquid in a nanotube-based continuum model. However, the specifics of the complexes, consisting of a finite number of molecules, should be kept in mind. These complexes, called clusters in literatures, are intermediately located between the bulk matter and individual particles(atoms or molecules). The fact of heterogeneity of water is now experimentally established [27].

There are groups of molecules in liquid as "microcrystals" containing tens or hundreds of molecules. Each microcrystal maintains solid form. These groups of molecules or "clusters"exist for a short period of time, then break up and are re-created again. Besides, they are constantly moving so that each molecule does not belong at all times to the same group of molecules, or "cluster."

Modelingpredicts that gas moleculesbounce off theperfectly smoothinner walls of thenanotubesasbilliard balls, andwater moleculesslide overthemwithout stopping. Possible cause of unusually rapid flow of water is maybe due to the small-diameter nanotube molecules move on them orderly, rarely colliding with each other. This "organized" move is much faster than usual chaotic flow. However, while the mechanism of flow of water and gas through the nanotubes is not very clear and only further experiments and calculations can helpunderstand it.

The model of mass transfer of liquid in a nanotube proposed in this research is based on the availability of nanoscale crystalline clusters in it [19].

A similar concept was developed by researchers in which the model of structured flow of fluid through the nanotube is considered. It is shown that the flow character in the nanotube depends on the relation between the equilibrium crystallite size and the diameter of the nanotube.

Figure 4.11 shows the resultsof calculationsby the molecular dynamicsof fluid flowin the nanotubein a plane (a)andthree-dimensional state (b).The figure showsthe ordered regionsof the liquid.

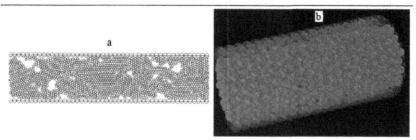

FIGURE 4.11   The results of calculations of fluid flow in the nanotube.

The typical size of crystallite is 1–2 nm, that is,compared, for example, with a diameter of silica nanotubes of different composition and structure.

The flow model proposed in the present work is based on the presence of "quasi-solid" phase in the central part of the nanotube and liquid layer, non-autonomous phases [23].

Consideration of such a structure that is formed when fluid flows through the nanotube, is also justified by the aforementioned results of the experimental studies and molecular modeling.

When considering the fluid flow with such structure through the nanotube, we will take into account the aspect ratio of "quasi-solid" phase and the diameter of the nanotube. Then the character of the flow is stable and the liquid phase can be regarded as a continuous medium with viscosity η.

Let's establish relationship between the volumetric flow rate of liquidflowing from a liquid layer of the nanotube length$l$, the radius R and the pressure drop $\Delta p / l$, $\Delta p = p - p_0$, where, $p_0$is the initial pressure in the tube (Fig. 4.12).

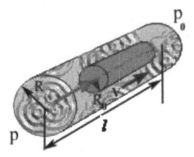

FIGURE 4.12   Flow through liquid layer of the nanotube.

Let R_0 be a radius of the tube from the "quasi-solid" phase and v – velocity of fluid flow through the nanotube.

Structural regime of fluid flow (Fig. 4.13) implies the existence of the continuous laminar layer of liquid (the liquid layer in the nanotube) along the walls of a pipe. In the central part of a pipe a core of the flow is observed, where the fluid moves, keeping his former structure, i.e. as a solid ("quasi-solid" phase in the nanotube). The velocity slip is indicated in Fig. 4.13a through v_0.

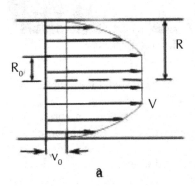
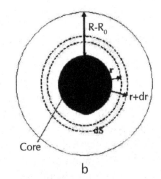

FIGURE 4.13   Structure of the flow in the nanotube.

Let's find the velocity profile $v(r)$ in a liquid interlayer $R_0 \leq r \leq R$ of the nanotube. We select a cylinder with radius r and length l in the interlayer, located symmetrically to the center line of the pipe (see Fig. 4.13).

At the steady flow, the sum of all forces acting on all the volumes of fluid with effective viscosity η, is zero.

The following forces are applied on the chosen cylinder: the pressure force and viscous friction force affects the side of the cylinder with radius r, calculated by the Newton formula.

Thus,

$$(p-p_0)\pi r^2 = -\eta\frac{dv}{dr}2\pi rl \tag{4.6}$$

Integrating Eq. (4.6) between r to R with the boundary conditions $r = R : v = v_0$, we obtained a formula to calculate the velocity of the liquid layers located at a distance r from the axis of the tube:

$$v(r) = (p-p_0)\frac{R^2 - r^2}{4\eta l} + v_0 \tag{4.7}$$

Maximum speed v_y has the core of the nanotube $0 \le r \le R_0$ and is equal to:

$$v_r = (p - p_0)\frac{R^2 - R_0^2}{4\eta l} + v_0 \tag{4.8}$$

Such structure of the liquid flow through nanotubes considering the slip is similar to a behavior of viscoplastic liquids in the tubes. For viscoplastic fluids a characteristic feature is that they are to achieve a certain critical internal shear stresses τ_0 and behave like solids. Meanwhile, when internal stress exceeds a critical value begin to move as normal fluid. In the work researchers the liquid behaves in the nanotube the similar way. A critical pressure drop is also needed to start the flow of liquid in a nanotube.

Structural regime of fluid flow requires existence of continuous laminar layer of liquid along the walls of pipe. In the central part of the pipe is observed flow with core radius R_{00}, where the fluid moves, keeping his former structure (i.e., as a solid).

The velocity distribution over the pipe section with radius R of laminar layer of viscoplastic fluid is expressed as follows:

$$v(r) = \frac{\Delta p}{4\eta l}(R^2 - r^2) - \frac{\tau_0}{\eta}(R - r) \tag{4.9}$$

The speed offlow corein $0 \le r \le R_{00}$ is equal

$$v_r = \frac{\Delta p}{4\eta l}(R^2 - R_{00}^2) - \frac{\tau_0}{\eta}(R - R_{00}) \tag{4.10}$$

Let's calculate the flow or quantity of fluid flowing through the nanotube cross-section S at a time unit. The liquid flow dQ for the inhomogeneous velocity field flowing from thecylindrical layer ofthickness dr, which is located at a distance r from the tube axis is determined from the relation,

$$dQ = v(r)dS = v(r)2\pi r dr \tag{4.11}$$

where dS–the area of the cross-section of cylindrical layer (between the dotted lines in Fig. 4.13).

Let's place Eq. (4.7) in Eq. (4.11), integrate over the radius of all sections from R_0 to R and take into account that the fluid flow through the core flow is determined from the relationship $Q_x = \pi R_0^2 v_x$. Then we get the formula for the flow of liquid from the nanotube:

$$v_0 \Delta p R^2 v_0 + Q_P \left[1 - \left(\frac{R_0}{R} \right)^4 \right]$$
(4.12)

If $(R_0 / R)^4 \ll$ (no nucleus) and $v_0 \Delta p R^2 / 8\eta$ (no slip), then Eq. (4.12) coincides with Poiseuille formula (the Eq. (3.5)). When $R_0 \approx R$ (no of a viscous liquid interlayer in the nanotube), the flow rate Q is equal to volumetric flow $Q \approx \pi R^2 v_0$ of fluid for a uniform field of velocity (full slip).

Accordingly, flow rate of the viscoplastic fluid flowing with a velocity (4.7), is equal to:

$$Q = -\frac{\pi R^3 \tau_0}{3\eta} \left[1 - \left(\frac{R_{00}}{R} \right)^3 \right] + Q_P \left[1 - \left(\frac{R_{00}}{R} \right)^4 \right]$$
(4.13)

Comparing the Eqs. (4.7), (4.8), (4.9), (4.10) and (4.12), (4.13), we can see that the structure of the flow of the liquid through the nanotubes considering the slippage, is similar to that of the flow of viscoplastic fluid in a pipe of the same radius R.

Given that the size of the central core flow of viscoplastic fluid (radius R_{00}) is defined by:

$$R_{00} = \frac{2\tau_0 l}{\Delta p}$$
(4.14)

for viscoplastic fluid flow we obtain Buckingham formula:

$$Q = Q_P \left[1 + \frac{1}{3} \left(\frac{2l\tau_0}{R\Delta p} \right)^4 - \frac{4}{3} \left(\frac{2l\tau_0}{R\Delta p} \right) \right]$$
(4.15)

We'll establish conformity of the pipe that implements the flow of a viscoplastic fluid with a fluid-filled nanotube, the same size and with the same pressure drop. We say that an effective internal critical shear stress τ_{0ef} of viscoplastic fluid flow, which ensures the coincidence rate with the flow of fluid in the nanotube. Then from Eq. (4.15) we obtain equation of fourth order to determine τ_{0ef}:

$$\left(\frac{2l\tau_{0ef}}{R\Delta p} \right)^4 - 4 \left(\frac{2l\tau_{0ef}}{R\Delta p} \right) = A, A = 3(\varepsilon - 1), \varepsilon = Q / Q_P$$
(4.16)

The solution of Eq. (4.16) can be found, for example, the iteration method of Newton:

$$\overline{\tau}_{0ef_n} = \overline{\tau}_{0ef_{n-1}} - \frac{\overline{\tau}_{0ef_{n-1}}^{-4} - 4\overline{\tau}_{0ef_{n-1}} - A}{4\overline{\tau}_{0ef_{n-1}}^{-3} - 4}, \overline{\tau}_{0ef} = \frac{2l\overline{\tau}_{0ef}}{R\Delta p} \qquad (4.17)$$

The first component in Eq. (4.12) represents the contribution to the fluid flow due to the slippage, and it becomes clear that the slippage significantly enhances the flow rate in the nanotube, when $l\eta v_0 > \approx \Delta pR^2$.

This result is consistent with experimental and theoretical results of Kalra et al. [34], Majumder et al. [41], Skoulidas et al. [66], Hummer et al. [28], and Holt et al. [25], which show that water flow in nanochannels can be much higher than under the same conditions, but for the liquid continuum.

In the absence of slippage $\varepsilon = 1$ the Eq. (4.16) has a trivial solution $\overline{\tau}_{0ef} = 0$.

4.3 THE RESULTS OF THE CALCULATIONS

Let's determine the dependence of the effective critical inner shear stress $\tau_{0ef} = 0$ on the radius of the nanotubes, by taking necessary values for calculations $\varepsilon = Q/Q_p$ from the work of Thomas John and McGaughey Alan [73]. The results of calculations at $\Delta p/l = 2.1 \times 10^{14} Pa/m$ are in the tablebelow:

TABLE 4.2   Results of the Calculations

R, м	τ_{0ef} (Па)	$\varepsilon = Q/Q_P$
0.83×10^{-9}	498,498	350
1.11×10^{-9}	577,500	200
1.385×10^{-9}	632,599	114
1.665×10^{-9}	699,300	84
1.94×10^{-9}	782,208	68
2.22×10^{-9}	855,477	57
2.495×10^{-9}	932,631	50

Calculations show that the value of effective internal shear stress depends on the size of the nanotube. Figure 4.14 shows the dependence τ_{0ef} on the nanotube radius, and Fig. 4.15 shows the structure of the flow.

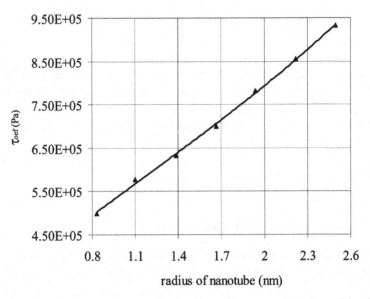

FIGURE 4.14 Dependence of the effective inner shear stress from the radius of the nanotube.

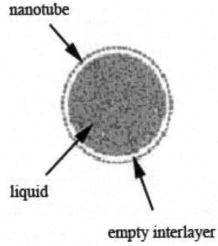

FIGURE 4.15 The structure of the flow.

The dependence τ_{0ef} (R) is almost linear. Within the range of the considered nanotube sizes τ_{0ef} has a relatively low value, which indicates the smoothness of the surface of carbon nanotubes.

4.4 THE FLOW OF FLUID WITH AN EMPTY INTERLAYER

The works of Kotsalis et al. [35] and Xi Chen et al. [80] were analyzed in the aforementioned analysis of the structure of liquid flow in carbon nanotubes. The results of the calculations of the cited works (Figs. 4.16 and 4.17) showed that during the flow of the liquid particles, an empty layer between the fluid and the nanotube is formed. The area near the walls of the carbon nanotube $R_* \leq r \leq R$ becomes inaccessible for the molecules of the liquid due to van der Waals repulsion forces of the heterogeneous particles of the carbon and water (Fig. 4.2). Moreover, according to the results of Kotsalis et al., [35] and Xi Chen et al. [80] thicknesses of thelayers $R_* \leq r \leq R$ regardless of radiuses of the nanotubes are practically identical: $R_* / R \approx 0{,}88$.

A similar result was obtained in Hongfei Ye et al.[26], which is an image Fig. 4.16 of the configuration of water molecules inside (8, 8) single-walled carbon nanotubes at different temperatures: 298, 325, and 350°K.

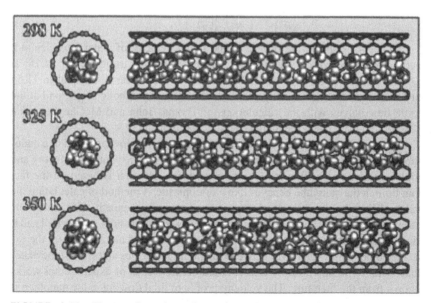

FIGURE 4.16 The configuration of water molecules inside single-walled carbon nanotubes.

Integrating Eq. (4.6) between r to R at the boundary conditions $r = R_*$:
$v = v_*$, we obtain a formula to calculate the velocity of the liquid layers located
at a distance r from the axis of the tube:

$$v(r) = v_* + \frac{2Q_P}{\pi R^2}\left[\left(\frac{R_*}{R}\right)^2 - \left(\frac{r}{R}\right)^2\right] \tag{4.18}$$

Let's insert Eq. (4.18) in Eq. (4.11), integrate over the radius of all sections from 0
to R_*. Then we get a formula for the flow of liquid from the nanotube:

$$Q = Q_P\left[v_* \frac{8l\eta R_*^2}{\Delta p R^4} + \left(\frac{R_*}{R}\right)^4\right] \tag{4.19}$$

or

$$\varepsilon = \frac{8l\eta v_*}{\Delta p R^4}\left(\frac{R_*}{R}\right)^2 + \left(\frac{R_*}{R}\right)^4$$

from which we can determine the unknown v_*:

$$v_* = \frac{Q_P}{\pi R^2}\left[\varepsilon - \left(\frac{R_*}{R}\right)^4\right]\left(\frac{R}{R_*}\right)^2 \tag{4.20}$$

Figure 4.17 shows the profile of the radial velocity of water particles in a
carbon nanotube with a diameter of 2.77 nm, calculated using the Eq. (4.18) at
$Q_P = 4,75 \times 10^{-19} n / m^2$, $\varepsilon = 114$. The velocity at the border v_* is equal to 11.55
m/s. It is seen that the velocity profile is similar to a parabolic shape, and at the
same time agrees with the calculations of Thomas John and McGaughey Alan
[73], obtained by using the model of molecular dynamics.

The calculations suggest the following conclusions. Flow of liquid in a nanotube was investigated using synthesis of the methods of the continuum theory and
molecular dynamics. Two models are considered. The first is based on the fact
that fluid in the nanotube behaves like a viscoplastic. A method of calculating the
value of limiting shear stress is proposed, which was dependent on the nanotube
radius. A simplified model agrees quite well with the results of the molecular
simulations of fluid flow in carbon nanotubes. The second model assumes the existence of an empty interlayer between the liquid molecules and wall of the nanotube. This formulation of the task is based on the results of experimental works
known from the literature. The velocity profile of fluid flowing in the nanotube is
practically identical to the profile determined by molecular modeling.

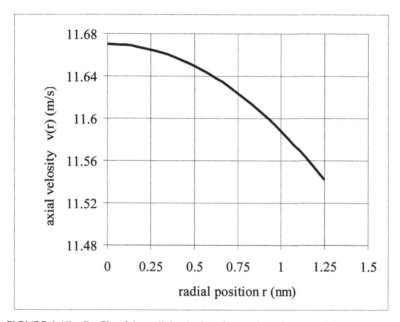

FIGURE 4.17   Profile of the radial velocity of water in carbon nanotubes.

As seen from the results of the calculations, the velocity value varies slightly along the radius of the nanotube. Such a velocity distribution of the fluid particles can be explained by the lack of friction between the molecules of the liquid and the wall due to the presence of an empty layer. This leads to an easy slippage of the liquid and, consequently, anomalous increase in flow compared to the Poiseuille flow.

KEYWORDS

- **Fluid Flow**
- **Interacting Particles**
- **Liquid**
- **Modeling**
- **Nanotube**
- **Poiseuille flow**

REFERENCES

1. Arie, A. G., &Slavkin, B. C. (1995). About the mechanism of oilgas saturation of sand lenses. *Oil and Gas Geology*, № 2.

2. Bakajin Olgica, Duke Thomas, A. J., Tegenfeldt Jonas, Chou Chia, Fu., Chan Shirley, S., & Austin Robert, H., Cox Edward, C., Separation of 100-Kilobase DNA Molecules in 10 Seconds, *Analytical Chemistry*; 73(24); 6053–6056 (2001).
3. Biomolecules with Microdevices", Electrophoresis, 21 (1), 81–90 (2000) Barrat, J. L., Chiaruttini, F., *Mol. Phys.* (2003)101, 1605–1610.
4. Bortov, V., Yu, Garanin, D. I., Georgievski, V., Yu. et.al. Comparative tests of reforming catalysts by "AKSENS". *Petrochemical and Refining*, (2003), vol. 2, p. 10–17.
5. Buyanov, R. A., Krivoruchko, O. P., Development of the theory of crystallization of sparingly soluble metal hydroxides and scientific basis of catalyst preparation of compounds of this class. *Kinetics and Catalysis*, (1976) v. 17, № 3, p. (765–775).
6. Cao, G., Qiao, Y., Zhou, Q., & Chen X. *Molecular Simulation* (2008), 88, 371–378.
7. Chernoivanov, V. I., Mazalov, Yu. A., Soloviev, R., Yu etc. A method of making the composition. RF Patent on application № 2003119564 from (02.07. 2003).
8. Chemical Encyclopedia, Ed., Knunyants, I. L., Moscow: Soviet Encyclopedia, (1990) vol. 1, 2.
9. Chou, C. F., Austin, R. H., Bakajin, O., Tegenfeldt, J. O., Castelino, J. A., Chan, S. S., Cox, E. C., Craighed, H., Darnton, N., Duke, T. A. J., Han, J., & Turner, S., Sorting biomolecules with microdevices, <http://www.ncbi.nlm.nih.gov/ pubmed/10634473##>Electrop horesis<http://www.ncbi.nlm.nih.gov/pubmed/10634473##>. (2000) 21(1):81–90.
10. Churaev, N. V., Ralston, J., Sergeeva, I. P., & Sobolev, V. D., Electrokinetic properties of methylated quartz capillaries. *Journal of Colloid and Interface Science*, (2002) V. 96, p. 265–278.
11. Cottin-Bizonne, C., Barentin, C., & Charlaix, E., Dynamics of simple liquids at heterogeneous surfaces: Molecular-dynamics simulations and hydrodynamic description. *The European Physical Journal E*, (2004), V. 15, p. 427–438.
12. Cottin-Bizonne, C., Cross, B., Steinberger, A., & Charlaix, E., Boundary Slip on Smooth Hydrophobic Surfaces: Intrinsic Effects and Possible Artifacts. *Physical Review Letters*, (2005), V. 94, p. 056102.
13. Culligan, Taewan Kim, & Yu Qiao, Nanoscale Fluid Transport: Size and Rate Effects. *NANO LETTERS*, Vol. 8, No. 9, (2008), p. 2988–2992.
14. Darrigol, O., Between hydrodynamics and elasticity theory: the first five births of the Navier-Stokes equations. *Archive for History of Exact Sciences*. (2002), V.56, p. 95–150.
15. Diskej, P. A., Possible Primary Migration of Oil from Source Rock in Oil Phase // *Bull. AAPG.* – (1975). – Vol. 59, № 2.
16. Ermolenko, N. F., & Efres, M. D., Regulation of the porous structure of oxide adsorbents and catalysts. Moscow: *Nauka*, (1991).
17. Evstrapov, A. A. The course of lectures "Nanotechnology in Environment and Medicine", (2011), 136 p.
18. Ficini, J., Lumbroso-Bader, N., & Depeze, J. K., Fundamentals of physical chemistry. – M., *Mir*, (1972).
19. Frenkel Ya, I., *UFN*, (1941), vol.25, no. 1, p. 1–18.
20. Friction laws at the nanoscale. *Nature*, 26.02.2009.
21. Gairik Sachdeva, Abhay Poal, Nelson Vadassery, Sayash Kumar, Srikar Chindada, Hanasaki, I., & Nakatani, A., *J., Chem. Phys.* (2006), 124, 144708.
22. Godymchuk, Yu, A., Ilyin, A. P., & Astankova, A. P., Oxidation of aluminum nanopowder in liquid water during heated. Proceedings of the Tomsk Polytechnic University, (2007), v. 310, № 1, p. 102–104.
23. Gusarov, V. V., Popov, I., Yu., & Il Nuovo Cimento, D., (1996), v.18, №7, pp. 799–805.
24. Hemanth Giri Rao, Will Water Flow Through Nanotubes? www.dstuns.iitm. ac.in/teaching-and-presentations/, (2010).

25. Holt, J. K., Park, H. G., Wang, Y., Stadermann, M., Artyukhin, A. B., Grigoropoulos, C. P., Noy, A., & Bakajin, O., Fast Mass Transport Through Sub-2nm Carbon Nanotubes. *Science*, (2006), v. 312, p. 1034–1037.
26. Hongfei Ye, Hongwu Zhang, Zhongqiang Zhang, & Yonggang Zheng. Size and temperature effects on the viscosity of water inside carbon nanotubes. *Nanoscale Research Letters* (2011), p. 6–87, http://www.nanoscalereslett. com/content/6/1/87.
27. Huang, C., Wikfeldt, K. T., Tokushima, T., Nordlund, D., Harada, Y., Bergmann, U., Niebuhr, M., Weiss, T. M., Horikawa, Y., Leetmaa, M., Ljungberg, M. P., Takahashi, O., Lenz, A., Ojamae, L., Lyubartsev, A. P., Shin, S., Petterson, L. G. M., & Nilsson, A., The inhogeneous structure of water at ambient conditions. *Proceeding of the National Academy of Sciences*; http:/www.pnas.org/content/early/2009/08/13/0904743106.abstract.
28. Hummer, G., Rasaiah, J. C., & Noworyta, J. P., Nature (2001), 414, 188–190.
29. Ilyin, A. P., Gromov, A. A., & Yablunovsky, G. V., About activity of aluminum powders. *Physics of Combustion and Explosion*, (2001), v. 37, № 4, p. 58–62.
30. Ilyin, A. P., Godymchuk, Yu, A., & Tikhonov, D. V., Threshold phenomena in the oxidation of aluminum nanopowders. Physics and chemistry of ultrafine (nano-) systems: Proc. VII All-Russian. Conf. Moscow: Moscow Engineering Physics Institute Printing, (2005), p. 178–179.
31. Jason, K. H., Hyung, G. P., Yinmin, W., Stadermann, M., Artyukhin, A. B., Grigoropoulos, C. P., Noy, A., & Bakajin, O., Fast Mass Transport Through Sub-2-Nanometer Carbon Nanotubes. *Science*, (2006). 312, 1034. pp. 1034–1037.
32. John, A., Thomas, & Alan,J. H.McGaughey, OttoleoKuter-Arnebeck. Pressure-driven water flow through carbon nanotubes: Insights from molecular dynamics simulation. *International Journal of Thermal Sciences*, 49, p. 281–289, (2010), journal homepage: www. elsevier.com/locate/ijts.
33. Joseph, P., & Tabeling, P., *Phys. ReV. E*, (2005), 71, 035303.
34. Kalra, A., Garde,S., & Hummer,G., Osmotic water transport through carbon nanotube arrays. *Proeedings of the National Academy of Sciences of the USA*, (2003), v. 100, p. 10175–10180.
35. Kotsalis, E. M., Walther, J. H., & Koumoutsakos, P., Multiphase water flow inside carbon nanotubes. *International Journal of Multiphase Flow*, 30, (2004), p. 995–1010.
36. Kozlova, E. G., Emelianov, Yu, I., & Krasiy, B. V., The new catalysts for reforming of gasoline with an octane rating of 96–98. *Catalysis in Industry*, (2003), № 6.
37. Korchagina, Yu, I., & Chetverikov, O. P., Methods of assessing the generation of hydrocarbons produce oil. M.: Nedra, (1983).
38. Lauga, E., Brenner, M. P., & Stone, H. A., Microfluidics: the no-slip boundary condition in Handbook of Experimental Fluid Dynamics. New York: Springer, (2006).
39. Lauga, E., & Stone, H. A., Effective slip in pressure-driven Stokes flow. *Journal of Fluid Mechanics*, (2003), V. 489, p. 55–77.
40. Li, T. D., Gao, J., Szoszkiewicz, R., Landman, U., & Riedo, E., *Phys. Rev. B*. (2007). 75. 115415.
41. Majumder, M., Chopra, N., Andrews, R., & Hinds, B., (2005), *Nature* 438, 44–44.
42. Mazalov, Yu, A., Patent RF № 2158396. *The method of burning metal fuels*. (2000).
43. Mirzadzhanzade Kh, A., Maharramov, A. M., Yusifzade, Kh, B., Shabanov, A. L., Nagiyev, F. B., Mammadzadeh, R. B., & Ramazanov, M. A., Study the influence of nanoparticles of iron and aluminum in the process of increasing the intensity of gas release and pressure for use in oil production. News of Baku University. *Science Series*, № 1, 2005, p. 5–13.
44. Mirzadzhanzade Kh, A., Maharramov, A. M., Nagiyev, F. B., & Ramazanov, M. A., Nanotechnology applications in the oil industry. Proceedings of the II-nd Scientific Conference "Nanotechnology-production (2005)", November 30–December 1, (2005) Fryazino., p. 47–52.

45. Mirzadzhanzade Kh, A., Maharramov, A. M., & Nagiyev, F. B., On the development of nanotechnology in the oil industry. *"Azerbaijan's Oil Industry,"* № 10, (2005), p. 51–65.
46. Mirzadzhanzade Kh, A., Bakhtizin, R. N., Nagiyev, F. B., & Mustafayev, A. A., Nano hydrodynamic effects on the base use of micro embryonic technology. *"Oil and Gas Business"*, Volume 3, (2005), p. 311–315.
47. Mirzadzhanzade Kh., A., Shahbazov, E. G., Shafiev, Sh., Sh., Nagiyev, F. B., Osmanov, B. A., & Mammadzadeh, R. B., Nanotechnology in the oil and gas production: research, implementation and results. Book of abstracts. Khazarneftgazyatag – (2006). International Scientific Conference on October 25–26, 2006, p. 47.
48. Morten, Bo, Lindholm Mikkelsen, Simon Eskild Jarlgaard, Peder Skafte-Pedersen. ExperimentalNanofluidics. Capillary filling of nanochannels. MIC – Department of Micro and Nanotechnology Technical University of Denmark, June 20th, (2005).
49. Nagiyev, F. B., Nonlinear oscillations of gas bubbles dissolved in the liquid. Izv.AN Az.SSR, Serf.-Tech and Math. *Science*, № 1, (1985) p. 136–140.
50. Nagiyev, F. B., Khabeev, N. S., Dynamics of soluble gas bubbles. Proceedings of the Academy of Sciences USSA, *Fluid and Gas Mechanics*, № 6, (1985), p. 52–59.
51. Nagiyev, F. B., & Mustafin.Kh., R., Using of high technologges in oil production. Intensification of oil production with aid of nanohydrodynamic effects usage. Collection of thesis International workshop "Socio-economic aspects of the energy corridor linking the Caspian Region with E. U." Baku, Azerbaijan, April,11–12th, 2007, p. 28–37.
52. Natsuki, T., Endo, M., & Tsuda H. *J. Appl. Phys.* 99 034311, (2006).
53. Natsuki, T., Hayashi, T., & Endo, M., *J. Appl. Phys.* 97 044307, 2005.
54. Navier, C. L. M. H., Memoire sur les lois du mouvement des fluids. Mémoires Académie des Sciences de l'Institut de France. (1823), v.1, p. 389–440.
55. Neimark, I. E., The main factors influencing the porous structure of hydroxide and oxide adsorbents. *Colloid Journal*, (1982), Volume 4, № 4, p. 780–783.
56. Nigmatulin, R. I., Fundamentals of mechanics of heterogeneous media. M., *"Nauka"*, (1978), 336 p.
57. Nigmatulin, R. I., Dynamics of multiphase media. Part I, *"Nauka"* M., (1987), 464 p.
58. Popov, I., Yu, Chivilikhin, S. A., & Gusarov, V. V., Model of the structured liquid through the nanotube: http://rusnanotech09.rusnanoforum.ru/Public/ Large Docs/theses/rus/poster/04/Chivilikhin.pdf.
59. Ou, J., Perot, J. B., & Rothstein, J. P., Laminar drag reduction in microchannels using ultrahydrophobic surfaces. *Physics of Fluids*, (2004), V. 16, p. 4635–4643.
60. Ou, J., & Perot, J. B., Drag Reduction and µ-PIV Measurements of the Flow Past Ultrahydrophobic Surfaces. *Physics of Fluids*, (2005), V. 17, p. 103606.
61. Press release on the website of the University of Wisconsin-Madison. Models present new view of nanoscale friction, 25.02.2009.
62. Proskurovskaya, L. T., Physical and chemical properties of electroexplosive ultrafine aluminum powders: Dis. Ph. D., Tomsk, (1988), 155 p.
63. Ramazanova, E. E., Shabanov, A. L., & Nagiyev, F. B., Perspectives of nanotechnology method applications for intensification oil-gas production. Collection of thesis International workshop "Electricity Generation and emission trading in South Eastern Europe". Sofia, Bulgaria, 21 September, (2007).
64. Rothstein, J. P., & McKinley, G. H., *J. Non-NewtonianFluidMech*, (1999), 86, 61–88.
65. Semwogerere, D., Morris,J. F., & Weeks,E. R., *J. FluidMech*. (2007), 581, 437–451.
66. Skoulidas, A. I., Ackerman, D. M., Johnson, J. K., & Sholl, D. S., *Phys. ReV. Lett.* (2002), 89, 185901.
67. Sokhan, V. P., Nicholson, D., Quirke, N. J., *Chem. Phys.* (2002), 117,8531–8539.

68. Sorokin, V. S., Variational method in the theory of convection. *Applied Mathematics and Mechanics.* Volume XVII, (1953), p. 39–48.

69. Stepin, B. D., & Tsvetkov, A. A., Inorganic Chemistry. Moscow: Higher School, (1994), 608 p.

70. Suetin, M. V., & Vakhrushev, A. V., Molecular dynamics simulation of adsorption and desorption of methane storage managed nanocapsules. "All-Russian Conference with international participation the Internet "From nanostructures, nanomaterials and nanotechnologies for nanotechnology", *Izhevsk*, 08/04/2009, p. 112.

71. Sunyaev, Z. I., Sunyaev, R. Z., & Safiyeva, R. Z., Oil dispersions systems. M.: *Chemistry*, (1990).

72. Tienchong Chang, Dominoes in Carbon Nanotubes. *Physical Review Letters*, 101, 175501, 24 October (2008).

73. Thomas John, A., & McGaughey Alan, J. H., Reassessing Fast Water Transport Through Carbon Nanotubes. *NANO LETTERS*, (2008), Vol. 8, No. 9, p. 2788–2793.

74. Thomas John, A., & McGaughey Alan, J. H., Water Flow in Carbon Nanotubes: Transition to Subcontinuum Transport. prl 102, *Physical Review Letters*, p. 184502–1–184502-4, (2009).

75. Uchic1 Michael, D., Dimiduk1 Dennis, M., Florando Jeffrey, N., & Nix William, D., Sample Dimensions Influence Strength and Crystal Plasticity. *Science* 13 August (2004): vol. 305 №. 5686, pp. 986–989.

76. Wang, C. Y., Ru, C. Q., & Mioduchowski, A., *Phys. Rev. B* 72, 075414, (2005).

77. Wang Q., & Varadan, V. K., *Int. J. Solids Struc.* 43. 254, (2006).

78. Wei-xian Zhang, Nanoscale iron particles for environmental remediation: An overview. *Journal of Nanoparticle Research* № 5, pp. 323–332, (2003).

79. Whitby, M., & Quirke, N., Fluid flow in carbon nanotubes and nanopipes. Chemistry Department, Imperial College, South Kensington, London SW7 2AZ, UK. *Naturenanotechnology*,www.nature.com/ naturenanotechnology, vol. 2, p. 87–94, February (2007).

80. Xi Chen, Guoxin Cao, Aijie Han, Venkata, K., Punyamurtula, Ling Liu, & Patricia, J.,

81. Yoon, J., Ru, C. Q., & Mioduchowski, A., *Compos. Sci. Technol.* 63, 1533, (2003).

82. Yoon, J., Ru, C. Q., & Mioduchowski, A., *J. Appl. Phys.* 93. 4801, (2003).

83. Yoon, J., Ru, C. Q., & Mioduchowski, A. *J. Composites* B 35, 87, (2004).

CHAPTER 5

NANOHYDROMECHANICS

CONTENTS

5.1 NANOPHENOMENON IN OIL PRODUCTION

Oil-saturated layers are porous materials with different pore sizes, pore channels and composition of rocks that define the features of interaction formation and injected fluids with the rock. Taking the mentioned into account we can conclude that the displacement of oil from oil fields in production wells is not a mechanical process of substitution of oil displacing it with water, but a complex physical-chemical process in which the decisive role is played by the phenomenon of ion exchange between reservoir and injected fluids with the rock, i.e. nanoscale phenomena.

The mechanism of displacement of oil in the reservoir and its recovery is largely determined by the molecular-surface processes occurring at phase interfaces (the rock-forming minerals–saturate the reservoir fluids and gases–displacing agents). Therefore, the problem of wettability is one of the major problems in oil and gas field of nanoscience.

As the clay is an ultra-system, a huge amount of research on regulation of the position of clay minerals in porous media with good reason can be attributed to nanoscience. To it also should be included the study of gas hydrates, a number of processes regulating the properties of the pumped oil and gas trapped water, water–oil preparation.

Filteringoilin reservoirs at a depth of13 km of hard rock is determined by the hydro dynamics oft he Darcy law. Of the Navier-Stokes equation

$$\rho\left(\frac{v}{t}+(\nabla\cdot v)v\right)=-\nabla p+\mu v \tag{5.1}$$

At very low Reynolds numbers it follows that we can neglect the inertial forces and simplify viscous friction:

$$-\nabla p-\mu\frac{v}{d^2}=0 \tag{5.2}$$

where, d– characteristic pore size. Equation (5.2) yields Darcy's law

$$v=-\frac{K}{\mu}\nabla p \tag{5.3}$$

where, K = d²– permeability of the medium, a– viscosity of the fluid.

Typical permeability values range from 5 mDto 500 mD. The permeability of coarse-grained sandstone is $10^{-8}-10^{-9}\,sm^2$, the permeability of dense sand stone around $10^{-2}\,sm^2$. Medium with the permeability 1 sP passes of fluidf low with a viscosity 1 s P at a pressure gradient $1aym/sm$.

During filtration oil fills and moves in the pores with the size of 1 mkm. In order to displace oil from the reservoir one should have a medium with density and viscosity of oil and the size of 1 mkm.

It seems that it is necessary to conduct experimental and theoretical studies of possible ways to obtain microbubbles nanoscale environments, to make the calculations and estimates of energy costs upon receipt of such media in different ways and their application to problems of oil production.

5.2 PETROLEUM COMPOSITION

Chemically, oil is a complexmixture of hydrocarbons (HC) andcarbon compounds. Itconsists of the followingelements:carbon (84–87%), hydrogen (12–14%), oxygen, nitrogen, sulfur (1–2%). The sulfur contentcan reach up to3–5%. Oils can contain the followingparts:a hydrocarbon, asvalto-resinous, porphyrins, sulfur and ash. Oilhas adissolvedgas thatis released whenit comes to theearth's surface.

The main partof petroleumhydro carbons are different in their composition, structure and properties, whichmay be ingaseous, liquidandsolid state. Depending on thestructure of the molecules they are classified intothree classes–paraffinic, naphthenicand aromatic.but a considerable proportionof oil ishydrocarbonsof mixedstructurecontainingstructuralelements of all threeabove-mentioned classes. The structure ofthe moleculesdetermines theirchemical and physical properties.

Carbonis characterized byits ability to formchains in whichthe atomsare connected in serieswith each other. In remaining connections hydrogen atoms are attached to the carbon. The number of carbon atoms in the molecules of paraffinic hydrocarbons exceeds the number of hydrogen atoms twice, with some constant excess in all the molecules equal to 2. In other words, the general formula of this classof hydrocarbons is C_nH_{2n+2}. Paraffinic hydrocarbonschemicallymore stableand referto the limiting HC.

Depending on the numberof carbon atomsin the molecule hydrocarbons maybe in one of the threestates of aggregation.

Thus, paraffin hydrocarbonsinoilcan be represented bygases, liquids and solidcrystalline substances. Theyhave different effectson the propertiesof oil:gasreduces viscosityandincreases the vapor pressure.

Fluidparaffinsdissolve wellinoilonly at elevated temperatures, forming a homogeneous mixture. Hardparaffins alsodissolve well inoilformingthe truemolecular mixtures. Paraffinhydrocarbons (with the exception of ceresin) can be easily crystallized inthe form of platesand platestrips.

Naphthenic(tsiklanovae oralicyclic) hydrocarbonshavecyclic structure (C/C_nH_{2n}), to be exact, they arecomposed of severalgroups–CH_2–interconnectedinringedsystem. Oil contains mainly naphthenes consisting off iveor six groups of CH_2.Allconnections of carbon and hydrogen are saturated, so then aphtheni coil has stableproperties.Compared withparaffin, naphthenes have a higherdensityandlowervapor pressureandhave bettersolvent power.

Aromatic hydrocarbons (arena) are represented by the formula C_nH_n, are most poor by hydrogen. The molecule has a form ofa ring with unsaturated carbon connections. The simplest representative of this class of hydrocarbons is benzene C_6H_6, which consists of six groups of CH_2. For aromatic hydrocarbons a large solubility, higher density and boiling point are typical.

Asphalt-resinous portion of oil is a substance of dark color, which is partially soluble in gasoline. They have the ability to swell in solvents, and then pass into mixture.

The solubility of asphaltenes in the resin-carbon systems increases with decreasing concentration of light hydrocarbons and increasing concentrations of aromatic hydrocarbons.

The resin does not dissolve in gasoline and is a polar substance with a relative molecular mass of 500–1200. They contain the bulk of oxygen, sulfur and nitrogen compounds of oil.

Asphaltic-resinous substances, and other polar components are surface-active compounds and natural oil-water emulsion stabilizers.

Special nitrogenous compounds of organic origin are called porphyrins. It is assumed that they were formed from animal hemoglobin and chlorophyll of plants. These compounds are destroyed at temperatures of 200–250°C.

Sulfur is prevalent in petroleum and hydrocarbon gas and is contained both in the free State and in the form of compounds (hydrogen sulfide, mercaptans).

Ash is the residue,which is formed by burning oil. This is a different mineral compound, usually iron, nickel, vanadium, and sometimes sodium.

Properties of oil determine the direction reprocessing and affect the products derived from petroleum, so there are different types of classification, which reflect the chemical nature of oil and determine possible areas of processing.

For example, in the base of the classification, reflecting the chemical composition is laid the preferencecontent of one or more classes of hydrocarbons in the oil.

Naphthene, paraffin, paraffin-naphthene, paraffin-naphthene-aromatic, naphthene-aromatic and aromatic hydrocarbons are being distinguished. Thus, all fractions in the paraffin oils contain a significant quantity of alkanes.

In the paraffin-naphthene-aromatic hydrocarbons of all three classes are contained in approximately equal amounts. Naphthene-aromatic oil is characterized mainly by the content of cycloalkanes and arenes, especially in the heavy fractions.

Classification is also used by the content of asphaltenes and resins.

the technical classification oils are divided into classes according to the sulfur content.

o   Types – by the output of factions at certain temperatures.

o   Groups – by the potential content of base oils.

o   Species – by content of solid alkanes (of papafins).

Figure 5.1 shows the components of the reservoir oil, which have different average integrated over the period of development and the entire volume of the reservoir values of physicochemical properties.

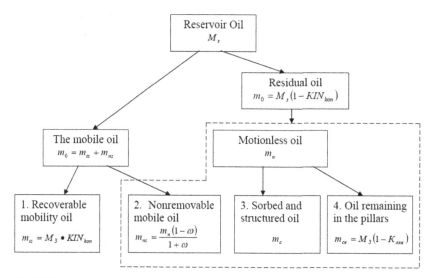

FIGURE 5.1 The constituents of reservoir oil.

Here, m_{u3} – reserves of reservoir oil at reservoir conditions; M_B– Mass of water in the reservoir area drained by the end of development, t; ω – watering at the end of development, the proportion of units; m_Π–Mass of mobile oil, t; m_{u3}– Mass of extracted oil, t; m_{H3}–Mass of the non-removable movable oil, non-removable t; m_c– Mass of the adsorbed and structured oil t; $m_{\mathcal{J}}$– Mass of oil remaining in pillars, t.

The mobile oil (m_Π) is part of the reservoir oil moving along layer due to the impact of external influences.

Recoverable mobile oil (m_{u3}) is certain part of the mobile oil, which can be extracted from the reservoir as a result of industrial activity with taking into account of economic and technological limitations.

Non-removable mobile oil (m_{H3}) is part of the mobile oil, which will not be extracted from the reservoir using the technologies as a result of industrial activity on the economic and technological constraints.

The residual oil (m_o) is part of the reservoir oil, located in the reservoir at the end of the displacement.

Motionless oil (m_H) is part of the reservoir oil, remaining motionless in the reservoir due to external influence.

Sorbed and structured oil (m_c) is part of motionless oil, retained near the surface of the collector by the intermolecular interaction.

Oil remaining in pillars ($m_{\mathcal{J}}$) is part of motionless oil, is not involved in the process of drainage.

Based on the proposed separation of produced oil into separate components, the average integral value of the physico-chemical properties of produced oil $\overline{X}_3$ for counting of reserves must be found from the expression:

$$\overline{X}_3 = \frac{X_{u3} \cdot m_{u3} + X_{H3} \cdot m_{H3} + X_c \cdot m_c + X_{U} \cdot m_{U}}{M_3} \; ; \tag{5.4}$$

$$M_3 = m_{u3} + m_{H3} + m_c + m_{U} \tag{5.5}$$

or

$$X_3 = X_{u3} d_{u3} + X_{H3} d_{H3} + X_c d_c + X_{U} d_{U} \tag{5.6}$$

where, X_{u3}, X_{H3}, X_c, X_{U} – the mean of the integral value of the corresponding component of reservoir oil; d_{u3}, d_{H3}, d_c, d_{U} : – the mass fraction of the corresponding component of reservoir oil

$$d_i = m_i / M_3 ,$$

where, m_i – The mass of i component of reservoir oil; m_3 – The mass of reservoir oil (geological reserves of oil reservoir).

Equations (5.4) and (5.6) allow us to calculate the average integral value of the physicochemical properties of reservoir oil by using a different source (relative or absolute).

The average integral value of the physicochemical properties of mobile oil for use in the calculation of oil displacement processes must be found out of the expression:

$$\overline{X}_n = \frac{X_{u3}.m_{u3}}{m_n} + \frac{X_{H3}.m_{H3}}{m_n} \; ; \tag{5.7}$$

or

$$\overline{X}_n = X_{u3}.c_{u3} + X_{H3}.c_{H3} \tag{5.8}$$

where, $X_{u3} c_{u3} = m_{H3} / m_n$: – property value and the mass fraction of component of recovered part of mobile oil, respectively; $X_{H3} c_{H3} = m_{H3} / m_n$: – Property value and the mass fraction of component of non-removable part of mobile oil, respectively.

In practice, there are a number of contradictions in the calculation methods of geological reserves of hydrocarbons and the calculation of oil recovery processes. For example, an anachronism in the method of counting, which is based on the volume (sealer) ratio of produced oil, because the value of reserves is obtained depending on the conditions of oil, since the magnitude of the volume ratio is dependent on the parameters of technology of oil preparation (the number of stages of oil separation and the temperature and pressure conditions).

The density is determined for each formation zone in the case of inhomogeneity of the properties of reservoir oil in the layer. In this case, during the separation of the extracted products on the commodity oil and associated gas only their masses will be dependent on the properties of reservoir oil and the parameters of ground preparation.

5.3 NANOBUBBLES OF GAS, OIL IN THE PORECHANNELS AND WATER

The average diameter of the channels of the porous medium d is easy to estimate using the well known relationship.

$$d = \frac{4m}{S}$$
(5.9)

where, m – Porosity ofthe medium,fraction of a unit;S– Specific surface (surface per a volume)

Clay Oil Gas Motherboard Thicknesses (OGMT) usually characterized by a porosity of 10–20%, and their specific surface of not less than $10^8 m^{-1}$. Consequently, the diameters of the pore channels of clay OGMT not exceed an average of 3 nm.

The minimum diameter of a gas bubble in a water medium can be calculated from the following assumptions:

The gas pressure p_g in a bubble caused by the action of two components: of the deposit pressure $P_{n\pi}$ and the pressure of surface tension P_σ.

$$p_g = P_{n\pi} + P_\sigma$$
(5.10)

orinaccordance with the lawof Laplace

$$p_g = P_{n\pi} + \frac{2\sigma}{R}$$
(5.11)

where, σ – Coefficient of the surface tension of the liquid, in which formed a gas bubble; R – the bubble radius [50, 57].

Gas bubbles can be considered an ensemble of hydrocarbon molecules with a mass equal to or less than the buoyant force acting on it in the liquid, which gives a formal record

$$NM_g = \rho W_g$$
(5.12)

where, N – Number of moleculesin the ensemble; M – Massof each molecule; g – Acceleration of free fall; ρ –Density of the liquid; W– Volumeof the bubble.

The molecular energy of the ensemble, keeping a bottle from collapse (i.e., from the dissolution of gas), is defined by the van der Waals forces, in which force of intermolecular interaction can be ignored, since the gas has virtually no internal pressure. In this case, the magnitude of this energy per unit volume of a gas bubble is equal to the pressure p_0 in the bubble:

$$p_0 = \frac{NkT}{W - W_0} \qquad (5.13)$$

where, T – Absolute temperature of thegas °K; k – Boltzmann's constant; $W_0 = Nw_0$; $w_0 = 12w$: – Volume per one molecule of gas at the critical temperature and pressure; W – the real volume ofagas molecule.

From (5.13) after simple transformations, withtaking into account expression-for w_0 we have

$$N = \frac{p_0 W}{kT + p_0 w_0} \qquad (5.14)$$

which after substitution inEq. (5.12) gives

$$p_0 = \frac{kT}{M / \rho - w_0} \qquad (5.15)$$

The condition of the bubbles floating necessarily implies the existence of the phase boundary, that is, surface bounding the volume of the bubble. A necessary and sufficient condition for such a boundary is the equality of pressures in gas and liquid phases:

$$p = -p_0 \qquad (5.16)$$

Substituting the values of pressures from Eqs. (5.11) and (5.15) gives the desired diameter of the bubble in the form of

$$d_n = 2R = \frac{4\sigma(Vp_0 - M / \rho)}{kT + p_{n\pi}(w_0 - M / \rho)} \qquad (5.17)$$

Calculation shows that the diameter of the molecule the simplest hydrocarbon gas–methane is equal to 0.38 nm, $w_0 = 0.345 \cdot 1^{-27} m^3$.

If thistake into account, that $M = 27.2 \cdot 10^{-27} kg$. $\rho \sim 10^3 kg / m^3$. $k = 1.38 \cdot 10^{-23} J /{}^0 K$. $T = 360^0 K$. $\sigma = 62.3 \cdot 10^{-3} n / m$, then from Eq. (5.17) is easy to determine the value of the minimum diameter of the gas bubble.

Under hydrostatic pressure of petroleum p_{nz} = 20 Mpa, it is 7 nm. This is more than twice the diameter of the pore channels. Consequently, in the cramped conditions of the pore space of OGMT formation of a gas bubble is impossible, because the capillary pressure in the ducts with a diameter of 3 nm to more than 5 times higher than the pressure p_0 at a given temperature. This violates the necessary condition for the existence of a gas bubble in a liquid (the Eq. (5.16)) and prevents phase separation, since the external pressure leads to a collapse of gas bubbles, that is, to its "dissolution" of the pore fluid, if it occurs at all.

It is clear that the required number of methane molecules to form a bubble is equal to

$$N = \frac{W_n}{w_0} = \frac{179,6 \cdot 10^{-27}}{0,345 \cdot 10^{-27}} = 520$$

where, W_n – Volume of the bubble with diameter of 7nm.

The minimum diameter of a bubble of oil in water is calculated based on approximately similar considerations.

A bubble of oil in an aqueous medium can be considered as an ensemble of molecules, which are able to form an interface between the liquid phases. The above (average) radius of the volume of such an ensemble can be obtained from the Eq. (5.11):

$$R = \frac{2\sigma}{p_g - p_{n\pi}} \tag{5.18}$$

where, σ – Border tension coefficient in the water- oil, equal to approximately $45 \cdot 10^{-3} n/m$: $p_g = p_{n1} + p.$,

where, P – additional molecular pressure in the equation of van der Waals forces, which for a liquid at condition of it incompressibility it is possible to express from the of the same equation of van der Waals forces:

$$p = -\left(\frac{kT}{w} + p_{n\pi} \right) \tag{5.19}$$

where, w – The real volume of one molecule of the liquid, and"minus" sign due to the fact that the pressure force directed toward the center of the bubble of oil.

Then from the Eqs. (5.18) and(5.19) we have finally

$$d_n = 2R = \frac{4\sigma w}{kT + wp_{n\pi}} \tag{5.20}$$

The resulting formula determines the minimum value of the diameter of the bubble of oil in a water medium. In this case the real volume of the hydrocarbon molecule with one carbon atom–methane, $w \sim 3,5 \cdot 10^{-28} m^3$.

The total mass of hydrocarbons that are brought by failution flow per unit area of contact OGMT–rock reservoir is defined as

$$G = \int_0^t V_G dt \tag{5.21}$$

where, V_G–velocity of hydrocarbon generation Organic Materials (OM) from OGMT; t–time; G–the productivity of generation.

The formation of hydrocarbons from the Dispersed Organic Material (DOM) can be regarded as an elimination process in which the initial component is the reactive part of the DOM, and the final product (hydrocarbon molecules). Then, the reaction velocity of elimination is written well-known formula

$$V = \varepsilon(\Gamma - x) = \varepsilon \Gamma e^{\varepsilon t} \qquad (5.22)$$

since $x = \Gamma\left(1 - e^{-\varepsilon t}\right)$.

In such a formulation, V–the velocity of formation hydrocarbon from DOM; ε–integrated constant of reaction velocity; Γ–the initial concentration of the reaction capable of part of the DOM in the breed; $\times$ – the mass of hydrocarbons that has developed in time t; ε – is usually determined experimentally from the velocity of formation of hydrocarbon out from the given DOM at different temperatures.

Equation (5.22) is valid for open systems in which the products of the reaction easily draw off the center of the reaction. In conditions of porous medium of natural OGMT with a limited pore volume it is necessary to introduce a factor considering the difficulty of derivatives removal (reaction products) into the Eq. (5.22).

As noted above, gas (or oil) cannot exist in the pore channels of clay medium in the form of separate phase, the elimination reaction can continue only until the capacity of the pore space is depleted relative to the hydrocarbon material.

The magnitude of this capacity is determined by the solubility of hydrocarbon in the pore water under specified pressure and temperature. The limiting value of the concentration of derivatives in the layer place express the maximum capacity of the reaction volume. The velocity of reaction considering this case, is limited by the introduction of Eq. (5.22) factor $\alpha = 1 - C/C_0$, where C–the current concentration of hydrocarbon in the pore water, C_0–its maximum value.

Thus, the velocity of generation of hydrocarbon per unit area of the roof of OGMT is (actually generation process is implemented in OGMT thick h, but all products of this generation pass through the surface of contact generating thickness with the collector. Therefore, the amount of generation is convenient to normalize the surface area of the roof OGMT.)

$$V_G = \varepsilon h \Gamma \left(1 - C/C_0\right) e^{-\varepsilon t} \qquad (5.23)$$

where, h–thickness of OGMT.

The magnitude of the current concentration of hydrocarbons in the flow of pore water, squeezed out from OGMT, by definition, is equal

$$C = \frac{x}{W} = \frac{h\Gamma}{W}\left(1 - e^{-\varepsilon t}\right) \qquad (5.24)$$

where, W– the volume of pore water, released as a result of compaction OGMT and/or passed through it over time t; x – the mass of the produced hydrocarbon substances.

Substituting Eqs. (5.21) and (5.23) (5.24), we have

$$G = \varepsilon h \Gamma \int_0^t \left[1 - \frac{h\Gamma\left(1 - e^{-\varepsilon t}\right)}{C_0 W} \right] e^{-\varepsilon t} dt \qquad (5.25)$$

which after integration and simple transformations gives

$$G = h\Gamma\left(1 - e^{-\varepsilon t}\right)\left[1 - \frac{h\Gamma}{2WC_0}\left(1 - e^{-\varepsilon t}\right) \right] \qquad (5.26)$$

According to the Eq. (5.23) obtained formula remains valid until $C < C_0$. When $C = C_0$ then the hydrocarbon generation ceases. If the current hydrocarbon concentration exceeds a specified limit (i.e., hydrocarbon products of degradation are beginning to stand out in a separate phase, forming gas bubbles or droplets of oil) then the formula loses physical meaning.

Equation (5.26) is valid during the primary migration of hydrocarbons and loses its physical meaning in the transition to the process of secondary migration. From this it follows that the inequality $G > 0$ is always performed, $G = 0$ at $t = 0$ and when $t \to \infty$, G tends to $G_0 = h\Gamma$, which is quite realistic.

Since $W = w_0 + W_\phi$

where, W_ϕ – volume of failuation, which flow through unit area of OGMT by thickness h, and w_0–volume of pore fluid contained in a block of OGMT with a single base and height h, value $h\Gamma / 2C_0 W$ is always less than 1. However, w_0 is the volume in which occurs in the reaction of degradation of DOM. It is equal to the volume of voids of OGMT, including the volume occupied by the DOM. This can be written as $W_0 = hn + h\Gamma / \rho$, which in turn suggests that the inequality

$$\frac{h\Gamma}{2C_0 W} = \frac{h\Gamma}{2C_0(hn + h\Gamma / \rho + W_\phi)} < 1$$

It should be noted that the density of DOM does not exceed the $2 / t / m^3$, and is oftensignificantlyless; $h / n > h\Gamma$ in most cases is of practical interest, and $2C_0$ at a depth of petroleum is always more than $1 kg/m^3$.

As formulated in the task, the movement of pore fluid in the clay thickness in the inviolate natural environment under the principles of geofluid dynamics of slow flow is realized in the fiylation regime. It is absolutely clear that the alien pore water molecules of hydrocarbon–derivatives will be forced to fall on the axis of the pore channels with greater frequency than the water molecules. This is due to the fact that near the walls of the pore channels molecules of water are additionally bonded by surface forces of mineral skeleton, whereas in the center of the channel the resultant of these forces is equal to zero [1]. Similar conclusion was obtained by Diskej [15]. Therefore, the concentration of hydrocarbons in failation flow is greater than C_0. However, the cramped conditions of the pore space in OGMT, as noted, do not allow such molecules to combine into a separate phase. Consequently, they are forced to migrate in a homogeneous unstable mixture with molecules of the pore water.

Based on the above, one can calculate the concentration of hydrocarbons in failation stream, if their concentration in the volume of pore space is equal to C. The ideology of this calculation is fairly obvious.

If the channel pore with radius R and the length l contains N molecules of a solvent such as water, we can write down

$$R^2 l = N.\frac{4}{3}\pi r^3$$

where, r– radius of molecules of the solvent, the total number of which is n_t at the fixed moment of time are located on the channel axis(i.e., there are always vacancies). Then $l = 2rn$, which after substituting into the initial equation gives

$$n_t = \frac{2}{3}N\frac{r^2}{R^2}$$

In other words, the number of molecules that fall at the same time on the axis of the pore channel, in R^2/r^2 time smaller than two-thirds of their total number N in the channel. And because the frequency of contact of the hydrocarbon molecules with the axis of the pore channel is prioritized compared with the water molecules, their concentration in this part of the pore space is equal to

$$C_\phi = C\frac{N}{n_t} = \frac{3}{2}C\frac{R^2}{r^2} \tag{5.27}$$

The resulting formula characterizes the concentration of hydrocarbon-substances in filation flow (the Eq. (5.27) holds for the reaction volumes commensurate with the volume occupied by the molecules of derivatives, which is characteristic of the pore space of clay OGMT).

A capillary pressure force directed against the forces of buoyancy occurs at the hydrocarbon–cluster contact with manifold overlapping tight layer and therefore;

$$K_{np} \le \frac{n_n}{2}\left[\frac{gH}{\sigma}(\rho_0 - \rho) + \sqrt{\frac{n}{2K}}\right]^2 \tag{5.28}$$

where, K_{np} and n_n – coefficients of permeability and porosity of the proposed tires, respectively; K and n– coefficients of permeability and porosity of the manifold, respectively; σ – border tension between the hydrocarbon-phase and the reservoir water;H –depth of reservoir;g–acceleration of free fall; ρ_0–densityof reservoir water; ρ – densityof the hydrocarbonphase.

From the known Arrhenius equation the coefficient of the reaction velocity of degradation is equal to

$$\varepsilon = Ae^{-E/RT}$$

(5.29)

where, E–the activation energyof degradation; R– Universal gas constant;T– Absolute temperatureof the system.

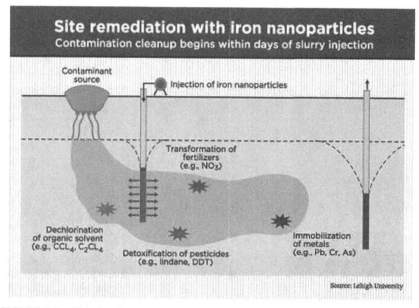

FIGURE 5.2   Cleaning the area with iron nanoparticles.

5.4   PROPERTIES OF ALUMINUM

Aluminum powder reacts with water at 50°C with an allocation of hydrogen. It reacts vigorously in exothermic reactions with oxygen-containing liquids with halogenated organic compounds and other oxidants.

Aluminum is relatively non-toxic, relatively inexpensive, widely distributed in nature and in large quantities produced in industry by the electrolysis.

Aluminumin the form ofnanopowderhaslowered the ability to reaction at room temperaturedue to the presenceof denseoxide-hydroxide shell, which is an electric doublelayer.

5.5   THE USE OF ALUMINUM POWDER IN OIL PRODUCTION

Aluminum powder isusedin the metallurgicalindustryinaluminothermy, as alloy-ingadditivesfor the manufacture ofsemi-finished productsbypressing and sinter-ing. Verystrongcomponents (gears, bushings, etc.) are received with this method. Powdersare alsousedinchemistry for thepreparation of compounds ofaluminumas a catalyst (e.g., production of ethyleneandacetone).

Considering high reactivity ofaluminum,especiallyin powder form, it is used in explosivesand solid fuelfor rockets, using itsproperty to quicklyignite.

In theoil industrynanoparticlesof aluminumwere usedfor the separation ofoil-water emulsion.

Figure 5.3 shows a snapshot of the device, where osmotic pressure was mea-sured. The pressure difference Δp was measured with a liquid manometer–U-shaped tube. Manometer was connected to a closed receptacle, which was par-tially vacuumed to avoid the effects of atmospheric pressure Δp.

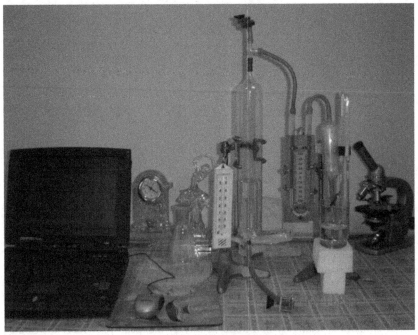

FIGURE 5.3   The picture of set, where measured the pressure.

The use of aluminum powder in injection wells for water injection leads to a significant increase in reservoir pressure, creating the effect of hydrobreak and facilitates the efficient displacement of oil.

KEYWORDS

- **Aluminum Powder**
- **Nanobubbles of Gas**
- **Nanophenomenon**
- **Oil Production**
- **Petroleum Composition**
- **Properties of Aluminum**

REFERENCES

1. Arie, A. G., & Slavkin, B. C. (1995). About the mechanism of oilgas saturation of sand lenses. *Oil and Gas Geology*, № 2.
2. Bakajin Olgica, Duke Thomas, A. J., Tegenfeldt Jonas, Chou Chia, Fu., Chan Shirley, S., & Austin Robert, H., Cox Edward, C., Separation of 100-Kilobase DNA Molecules in 10 Seconds, *Analytical Chemistry*; 73(24); 6053–6056 (2001).
3. Biomolecules with Microdevices", Electrophoresis, 21 (1), 81–90 (2000) Barrat, J. L., Chiaruttini, F., *Mol. Phys. (*2003)101, 1605–1610.
4. Bortov, V., Yu, Garanin, D. I., Georgievski, V., Yu. et.al. Comparative tests of reforming catalysts by "AKSENS". *Petrochemical and Refining*, (2003), vol. 2, p. 10–17.
5. Buyanov, R. A., Krivoruchko, O. P., Development of the theory of crystallization of sparingly soluble metal hydroxides and scientific basis of catalyst preparation of compounds of this class. *Kinetics and Catalysis*, (1976) v. 17, № 3, p. (765–775).
6. Cao, G., Qiao, Y., Zhou, Q., & Chen X. *Molecular Simulation* (2008), 88, 371–378.
7. Chernoivanov, V. I., Mazalov, Yu. A., Soloviev, R.,Yu etc. A method of making the composition. RF Patent on application № 2003119564 from (02.07. 2003).
8. Chemical Encyclopedia, Ed., Knunyants, I. L., Moscow: Soviet Encyclopedia, (1990) vol. 1, 2.
9. Chou, C. F., Austin, R. H., Bakajin, O., Tegenfeldt, J. O., Castelino, J. A., Chan, S. S., Cox, E. C., Craighed, H., Darnton, N., Duke, T. A. J., Han, J., & Turner, S., Sorting biomolecules with microdevices, <http://www.ncbi.nlm.nih.gov/ pubmed/10634473##> Electrophoresis <http://www.ncbi.nlm.nih.gov/pubmed/10634473##>. (2000) 21(1):81–90.
10. Churaev, N. V., Ralston, J., Sergeeva, I. P., & Sobolev, V. D.,Electrokinetic properties of methylated quartz capillaries. *Journal of Colloid and Interface Science*, (2002) V. 96, p. 265–278.
11. Cottin-Bizonne, C., Barentin, C., & Charlaix, E., Dynamics of simple liquids at heterogeneous surfaces: Molecular-dynamics simulations and hydrodynamic description. *The European Physical Journal E*, (2004), V. 15, p. 427–438.
12. Cottin-Bizonne, C., Cross, B., Steinberger, A., & Charlaix, E., Boundary Slip on Smooth Hydrophobic Surfaces: Intrinsic Effects and Possible Artifacts. *Physical Review Letters*, (2005), V. 94, p. 056102.
13. Culligan, Taewan Kim, & Yu Qiao, Nanoscale Fluid Transport: Size and Rate Effects. *NANO LETTERS*, Vol. 8, No. 9, (2008), p. 2988–2992.
14. Darrigol, O.,Between hydrodynamics and elasticity theory: the first five births of the Navier-Stokes equations. *Archive for History of Exact Sciences*. (2002), V.56, p. 95–150.

15. Diskej, P. A., Possible Primary Migration of Oil from Source Rock in Oil Phase // *Bull. AAPG.* – (1975). – Vol. 59, № 2.
16. Ermolenko, N. F., & Efres, M. D., Regulation of the porous structure of oxide adsorbents and catalysts. Moscow: *Nauka*, (1991).
17. Evstrapov, A. A. The course of lectures "Nanotechnology in Environment and Medicine", (2011), 136 p.
18. Ficini, J., Lumbroso-Bader, N., & Depeze, J. K., Fundamentals of physical chemistry. – M., *Mir*, (1972).
19. Frenkel Ya, I., *UFN*, (1941), vol.25, no. 1, p. 1–18.
20. Friction laws at the nanoscale. *Nature*, 26.02.2009.
21. Gairik Sachdeva, Abhay Poal, Nelson Vadassery, Sayash Kumar, Srikar Chindada, Hanasaki, I., &Nakatani, A., *J., Chem. Phys.* (2006), 124, 144708.
22. Godymchuk, Yu, A., Ilyin, A. P., & Astankova, A. P., Oxidation of aluminum nanopowder in liquid water during heated. Proceedings of the Tomsk Polytechnic University, (2007), v. 310, № 1, p. 102–104.
23. Gusarov, V. V., Popov, I., Yu., & Il Nuovo Cimento, D., (1996), v.18, №7, pp. 799–805.
24. Hemanth Giri Rao, Will Water Flow Through Nanotubes? www.dstuns.iitm. ac.in/teaching-and-presentations/, (2010).
25. Holt, J. K., Park, H. G., Wang, Y., Stadermann, M., Artyukhin, A. B., Grigoropoulos, C. P., Noy, A., & Bakajin, O.,Fast Mass Transport Through Sub-2nm Carbon Nanotubes. *Science*, (2006), v. 312, p. 1034–1037.
26. Hongfei Ye, Hongwu Zhang, Zhongqiang Zhang, & Yonggang Zheng. Size and temperature effects on the viscosity of water inside carbon nanotubes. *Nanoscale Research Letters* (2011), p. 6–87, http://www.nanoscalereslett. com/content/6/1/87.
27. Huang, C., Wikfeldt, K. T., Tokushima, T., Nordlund, D., Harada, Y., Bergmann, U., Niebuhr, M., Weiss, T. M., Horikawa, Y., Leetmaa, M., Ljungberg, M. P., Takahashi, O., Lenz, A., Ojamae, L., Lyubartsev, A. P., Shin, S., Petterson, L. G. M., & Nilsson, A., The inhogeneous structure of water at ambient conditions. *Proceeding of the National Academy of Sciences*; http:/www.pnas.org/content/early/2009/08/13/0904743106.abstract.
28. Hummer, G., Rasaiah, J. C., & Noworyta, J. P., Nature (2001), 414, 188–190.
29. Ilyin, A. P., Gromov, A. A., & Yablunovsky, G. V., About activity of aluminum powders. *Physics of Combustion and Explosion*, (2001), v. 37, № 4, p. 58–62.
30. Ilyin, A. P., Godymchuk, Yu, A., & Tikhonov, D. V., Threshold phenomena in the oxidation of aluminum nanopowders. Physics and chemistry of ultrafine (nano-) systems: Proc. VII All-Russian. Conf. Moscow: Moscow Engineering Physics Institute Printing, (2005), p. 178–179.
31. Jason, K. H., Hyung, G. P., Yinmin, W., Stadermann, M., Artyukhin, A. B., Grigoropoulos, C. P., Noy, A., & Bakajin, O., Fast Mass Transport Through Sub-2-Nanometer Carbon Nanotubes. *Science*, (2006). 312, 1034. pp. 1034–1037.
32. John,A., Thomas, &Alan,J. H. McGaughey, OttoleoKuter-Arnebeck. Pressure-driven water flow through carbon nanotubes: Insights from molecular dynamics simulation. *International Journal of Thermal Sciences*, 49, p. 281–289, (2010), journal homepage: www.elsevier.com/locate/ijts.
33. Joseph, P., & Tabeling, P., *Phys. ReV. E*, (2005), 71, 035303.
34. Kalra,A., Garde,S., &Hummer,G., Osmotic water transport through carbon nanotube arrays. *Proeedings of the National Academy of Sciences of the USA*, (2003), v. 100, p. 10175–10180.
35. Kotsalis, E. M., Walther, J. H., & Koumoutsakos, P., Multiphase water flow inside carbon nanotubes. *International Journal of Multiphase Flow*, 30, (2004), p. 995–1010.

36. Kozlova, E. G., Emelianov, Yu, I., & Krasiy, B. V., The new catalysts for reforming of gasoline with an octane rating of 96–98. *Catalysis in Industry*, (2003), № 6.
37. Korchagina, Yu, I., & Chetverikov, O. P., Methods of assessing the generation of hydrocarbons produce oil. M.: Nedra, (1983).
38. Lauga, E., Brenner, M. P., & Stone, H. A., Microfluidics: the no-slip boundary condition in Handbook of Experimental Fluid Dynamics. New York: Springer, (2006).
39. Lauga, E., & Stone, H. A.,Effective slip in pressure-driven Stokes flow. *Journal of Fluid Mechanics*, (2003), V. 489, p. 55–77.
40. Li, T. D., Gao, J., Szoszkiewicz, R., Landman, U., & Riedo, E., *Phys. Rev. B.* (2007). 75. 115415.
41. Majumder, M., Chopra, N., Andrews, R.,& Hinds, B., (2005), *Nature* 438, 44–44.
42. Mazalov, Yu, A., Patent RF № 2158396. *The method of burning metal fuels.* (2000).
43. Mirzadzhanzade Kh, A., Maharramov, A. M., Yusifzade, Kh, B., Shabanov, A. L., Nagiyev, F. B., Mammadzadeh, R. B., & Ramazanov, M. A., Study the influence of nanoparticles of iron and aluminum in the process of increasing the intensity of gas release and pressure for use in oil production. News of Baku University. *Science Series*, № 1, 2005, p. 5–13.
44. Mirzadzhanzade Kh, A., Maharramov, A. M., Nagiyev, F. B., & Ramazanov, M. A., Nanotechnology applications in the oil industry. Proceedings of the II-nd Scientific Conference "Nanotechnology-production (2005)", November 30–December 1, (2005) Fryazino., p. 47–52.
45. Mirzadzhanzade Kh, A., Maharramov, A. M., & Nagiyev, F. B., On the development of nanotechnology in the oil industry. *"Azerbaijan's Oil Industry,"* № 10, (2005), p. 51–65.
46. Mirzadzhanzade Kh, A., Bakhtizin, R. N., Nagiyev, F. B., & Mustafayev, A. A., Nano hydrodynamic effects on the base use of micro embryonic technology. *"Oil and Gas Business"*, Volume 3, (2005), p. 311–315.
47. Mirzadzhanzade Kh, A., Shahbazov, E. G., Shafiev, Sh., Sh., Nagiyev, F. B., Osmanov, B. A., & Mammadzadeh, R. B., Nanotechnology in the oil and gas production: research, implementation and results. Book of abstracts. Khazarneftgazyatag – (2006). International Scientific Conference on October 25–26, 2006, p. 47.
48. Morten, Bo, Lindholm Mikkelsen, Simon Eskild Jarlgaard, Peder Skafte-Pedersen. ExperimentalNanofluidics. Capillary filling of nanochannels. MIC – Department of Micro and Nanotechnology Technical University of Denmark, June 20th, (2005).
49. Nagiyev, F. B., Nonlinear oscillations of gas bubbles dissolved in the liquid. Izv.AN Az.SSR, Serf.-Tech and Math. *Science*, № 1, (1985) p. 136–140.
50. Nagiyev, F. B., Khabeev, N. S., Dynamics of soluble gas bubbles. Proceedings of the Academy of Sciences USSA, *Fluid and Gas Mechanics*, № 6, (1985), p. 52–59.
51. Nagiyev, F. B., &Mustafin.Kh., R.,Using of high technologges in oil production. Intensification of oil production with aid of nanohydrodynamic effects usage. Collection of thesis International workshop "Socio-economic aspects of the energy corridor linking the Caspian Region with E. U." Baku, Azerbaijan, April,11–12th, 2007, p. 28–37.
52. Natsuki, T., Endo, M., & Tsuda H. *J. Appl. Phys.* 99 034311, (2006).
53. Natsuki, T., Hayashi, T., & Endo, M., *J. Appl. Phys.* 97 044307, 2005.
54. Navier, C. L. M. H., Memoire sur les lois du mouvement des fluids. Mémoires Académie des Sciences de l'Institut de France. (1823), v.1, p. 389–440.
55. Neimark, I. E., The main factors influencing the porous structure of hydroxide and oxide adsorbents. *Colloid Journal*, (1982), Volume 4, № 4, p. 780–783.
56. Nigmatulin, R. I., Fundamentals of mechanics of heterogeneous media. M., *"Nauka"*, (1978), 336 p.
57. Nigmatulin, R. I., Dynamics of multiphase media. Part I, *"Nauka"* M., (1987), 464 p.

58. Popov, I., Yu, Chivilikhin, S. A., & Gusarov, V. V., Model of the structured liquid through the nanotube: http://rusnanotech09.rusnanoforum.ru/Public/ Large Docs/theses/rus/poster/04/Chivilikhin.pdf.

59. Ou, J., Perot, J. B., & Rothstein, J. P.,Laminar drag reduction in microchannels using ultra-hydrophobic surfaces. *Physics of Fluids*, (2004), V. 16, p. 4635–4643.

60. Ou, J., &Perot, J. B.,Drag Reduction and μ-PIV Measurements of the Flow Past Ultrahydrophobic Surfaces. *Physics of Fluids*, (2005), V. 17, p. 103606.

61. Press release on the website of the University of Wisconsin-Madison. Models present new view of nanoscale friction, 25.02.2009.

62. Proskurovskaya, L. T., Physical and chemical properties of electroexplosive ultrafine aluminum powders: Dis. Ph. D., Tomsk, (1988), 155 p.

63. Ramazanova, E. E., Shabanov, A. L., & Nagiyev, F. B., Perspectives of nanotechnology method applications for intensification oil-gas production. Collection of thesis International workshop "Electricity Generation and emission trading in South Eastern Europe". Sofia, Bulgaria, 21 September, (2007).

64. Rothstein, J. P., &McKinley, G. H., *J. Non-NewtonianFluidMech,* (1999), 86, 61–88.

65. Semwogerere,D., Morris,J. F., &Weeks,E. R., *J. FluidMech.* (2007), 581, 437–451.

66. Skoulidas, A. I., Ackerman, D. M., Johnson, J. K.,& Sholl, D. S., *Phys. ReV. Lett.* (2002), 89, 185901.

67. Sokhan, V. P., Nicholson, D., Quirke, N. J., *Chem. Phys.* (2002), 117,8531–8539.

68. Sorokin, V. S., Variational method in the theory of convection. *Applied Mathematics and Mechanics.* Volume XVII, (1953), p. 39–48.

69. Stepin, B. D., & Tsvetkov, A. A., Inorganic Chemistry. Moscow: Higher School, (1994), 608 p.

70. Suetin, M. V., & Vakhrushev, A. V., Molecular dynamics simulation of adsorption and desorption of methane storage managed nanocapsules. "All-Russian Conference with international participation the Internet "From nanostructures, nanomaterials and nanotechnologies for nanotechnology ", *Izhevsk*, 08/04/2009, p. 112.

71. Sunyaev, Z. I., Sunyaev, R. Z., & Safiyeva, R. Z., Oil dispersions systems. M.: *Chemistry*, (1990).

72. Tienchong Chang, Dominoes in Carbon Nanotubes. *Physical Review Letters*, 101, 175501, 24 October (2008).

73. Thomas John, A., & McGaughey Alan, J. H., Reassessing Fast Water Transport Through Carbon Nanotubes. *NANO LETTERS*, (2008), Vol. 8, No. 9, p. 2788–2793.

74. Thomas John, A., & McGaughey Alan, J. H., Water Flow in Carbon Nanotubes: Transition to Subcontinuum Transport. prl 102, *Physical Review Letters*, p. 184502–1–184502-4, (2009).

75. Uchic1 Michael, D., Dimiduk1 Dennis, M., Florando Jeffrey, N., & Nix William, D., Sample Dimensions Influence Strength and Crystal Plasticity. *Science* 13 August (2004): vol. 305 №. 5686, pp. 986–989.

76. Wang, C. Y., Ru, C. Q., & Mioduchowski, A., *Phys. Rev. B* 72, 075414, (2005).

77. Wang Q., & Varadan, V. K., *Int. J. Solids Struc.* 43. 254, (2006).

78. Wei-xian Zhang, Nanoscale iron particles for environmental remediation: An overview. *Journal of Nanoparticle Research* № 5, pp. 323–332, (2003).

79. Whitby, M., & Quirke, N., Fluid flow in carbon nanotubes and nanopipes. Chemistry Department, Imperial College, South Kensington, London SW7 2AZ, UK. *Naturenanotechnology*,www.nature.com/ naturenanotechnology, vol. 2, p. 87–94, February (2007).

80. Xi Chen, Guoxin Cao, Aijie Han, Venkata, K., Punyamurtula, Ling Liu, & Patricia, J.,

81. Yoon, J., Ru, C. Q., & Mioduchowski, A., *Compos. Sci. Technol.* 63, 1533 (2003).

82. Yoon, J., Ru, C. Q., & Mioduchowski, A., *J. Appl. Phys.* 93. 4801 (2003).
83. Yoon, J., Ru, C. Q., & Mioduchowski, A. *J. Composites* B 35, 87 (2004).

INDEX

Printed in the United States
by Baker & Taylor Publisher Services